Farmscape is evidence (once again) that Mary Swander's green thumb works its wonders outside of her bountiful Iowa garden. The play is a literary cornucopia, filled with stories that expose the soul of rural America. Rooted in a specific time and place, and yet as timeless as the *Iliad*. *Farmscape* is destined to become an American classic.

> —Osha Gray Davidson, *Broken Heartland: The Rise of America's Rural Ghetto*

Farmscape is a beautiful, often heart-breaking portrayal of the wrenching changes in Midwestern farming, told in the words of real people who live with this change. Some are surviving, some not. Some are accepting, some bitter. All are authentic and Mary Swander and her students have captured brilliantly the reality of their lives and turned them into a theater narrative that rings true with every word. There are few heroes here nor villains. But for everyone who thinks the idealized family farm still exists, *Farmscape* literally tells it like it is.

> —Richard Longworth, senior fellow at the Chicago Council on Global Affairs and author of *Caught in the Middle: America's Heartland in the Age of Globalism*

An inventive and intimate look at the lives of people throughout the spectrum of agriculture. The layering of voices allows for a sort of conversation that reflects how complex and complicated our discussion about farming really is.

> —Lisa Hamilton, *Deeply Rooted: Unconventional Farmers in the Age of Agribusiness*

Farmscape nails the rural experience. The play and commentaries reflect upon and push against, the "progress" we've made since the Farm Crisis of the 1980s. But why should we care what happens on farms in Iowa? Because what we do here has resonances for the rest of the agricultural world. Mary Swander and her students cast a broad net and the interviews they conducted create a classic piece of theatre, one that should be viewed by anyone wanting a better understanding of our complex and contradictory food system.

—Denise O'Brien, farmer and founder of Women, Food, and Agricultural Network

Mary Swander and her students have created a special thing—a striking piece of verbatim theatre. It's a rare treat to find a play that captures the agricultural experience. *Farmscape* brings to the stage a chorus of voices who have weathered the Farm Crisis and found their places in the changing rural environment. The accompanying essays detail how to produce the play in small communities, and how the issues raised by the drama reach beyond the regional to find global significance.

—John Peterson, lifelong farmer, Angelic Organics and star of *The Real Dirt on Farmer John*

Something that approaches the real diversity found in rural America, with impact far beyond the student playwrights' wildest dreams.

—*Farm Journal*

The fertile soil of Iowa is growing a vibrant new organism, Mary Swander's *Farmscape*. This wonderful project started with interviews by Swander's students with various Iowans about their lives on and off the farm. These were shaped by her poet's hand into a play, verbatim theater, then performed by townspeople in local readings in towns all over the Midwest, farmers, neighbors, interested parties, with "talk-back's" after those performances by engaged audiences. This is theater that really matters to everyone involved, and people in communities far beyond Iowa want in, want to participate in this unique experience themselves.

And now we have the play and essays by participants themselves and a host of gifted writers who know their way around agrarian themes. This wholesome process is bearing fruit and foretells a possible satisfying future, bringing art, empathy, and all kinds of people together to deepen connections and understanding. Mary Swander's students fanned out to find farmers' stories and she's bringing it all back home for us to be part of.

—Deborah Koons Garcia, film producer,
The Future of Food

Farmscape play at food conference evokes talk of today's agriculture.

—*Des Moines Register*

Farmscape

The Changing Rural Environment

Ice Cube Press
North Liberty, Iowa

Farmscape—
The Changing Rural Environment

Ice Cube Press, LLC (Est. 1993)
205 N. Front Street
North Liberty, Ia 52317
www.icecubepress.com
steve@icecubepress.com
@icecubepress on twitter

The paper used in this publication meets the minimum requirements of
the American National Standard for Information Sciences—Permanence
of Paper for Printed Library Materials, ANSI Z39.48-1992

Manufactured in the United States of America with recycled paper.

The *Farmscape* script contains some graphic and adult language.

Cover and interior photos: Special thanks to the Ames Historical Society,
David Perez Gomez, and Mary Swander for the photos used for this book.

Many thanks to my terrific students who worked with me and followed through on this project. And special appreciation to Fred Kirschenmann and Laura Miller at the Leopold Center for Sustainable Agriculture at Iowa State University who supported *Farmscape* from the beginning.

Contents—

Foreword
Agriculture and the Arts:
Terms of Endearment

Frederick Kirschenmann

In our modern culture we have succeeded in largely reducing agriculture to a set of scientific and technological questions. How can we make agriculture more efficient? What technologies can we invent to increase yields? How can we feed nine billion people?

Important as these questions may be, they often neglect to address some of the social, ethical, and ecological concerns which lie behind them—concerns that need to be entertained if we are to invent an agriculture and food system that is truly sustainable. Furthermore, we have created a modern society that largely separates most people from any experience with growing and producing food. So, we have mostly created a culture of separation.

It is highly unlikely that we will ever have a viable food and agriculture system on our landscapes if we fail to address these deeper social, ethical, and ecological concerns. Enter the Arts! A marriage of agriculture and the arts now becomes imperative to not only inspire, but also speak to the issues. As Gene Logsdon puts it in his visionary 2007 book, *The Mother of All Arts: Agrarianism and the Creative Impulse* (University of Kentucky Press): "That is what art and agriculture coming together could achieve: a sweet marriage of man and nature, of art and agriculture, of earth and paradise."

Logsdon is not alone in this observation. Wendell Berry's many poems and essays are clear demonstrations of the importance of entertaining the kind of philosophical and ethical questions we need to address if we are to meet the challenges ahead of us. Berry and others have demonstrated that we can foster these conversations if we marry agriculture and the arts. Liberty Hyde Baily, Dean of the New York State College of Agriculture at Cornell University (1903 -1913) who deliberately married science and art in his writings, stimulated the same kind of deep thinking. A horticulturist, poet, and journalist, Bailey wrote sixty-five books attempting to explain botany to lay people, and advocating, among other things, nature study and the appreciation of the countryside. Other luminaries, many of whom are featured in Logsdon's book, have similarly managed to merge agriculture and the arts to the same end.

And now we have Mary Swander and *Farmscape*. The drama consists of a collection of Iowa farmers' stories gathered by a class of students in Swander's Iowa State University English 557 course: Writing About Environmental Issues. To capture current agricultural challenges and opportunities, Swander gave her students a class assignment to go out and visit with some of the many, diverse farmers throughout the state and collect their stories in their own words. Swander and the students then arranged these stories into a readers' theater production that is now being presented in community theaters, colleges, and universities.

The result has been rather remarkable. *Farmscape* has now been presented throughout Iowa as well as communities in other states.

Invariably, hearing the farmers own stories in their own words presented in the context of an engaging and imaginative drama, invites exceptional conversations following the production of the play.

There are now at least three types of farm markets in Iowa and most other states. First, there are direct-market-farmers who mostly grow food products that they sell directly to their customers through farmers markets, CSA's, internet sales, and other venues. Secondly, we have undifferentiated-commodity-market-farmers who mostly produce one or two bulk commodities in massive quantities and sell them into food chains that turn those commodities into a stunning variety of food products. More recently we have seen the development of numerous marketing networks of (mostly mid-scale) farmers who produce a differentiated branded product that provides consumers with additional choices in the marketplace. Some examples of this third, aggregated market are Niman Ranch, Organic Valley Family of Farms, Shepherd's Grain, Country Natural Beef, Red Tomato, Eden Farms Berkshire Pork, and others.

All too often these different farms are either vilified or celebrated (by both farmers and consumers) depending on deeply held perspectives. Through the characters' own stories, *Farmscape* helps theatergoers better understand and appreciate the struggles and opportunities of each type of farming. The talkback, or conversation that follows the play, consequently produces a deeper knowledge and awareness of the variety of farmers and food choices that most citizens now have available to them. And, perhaps more importantly, the theatrical experience enhances everyone's grasp of everything farmers do everyday to provide the food we all eat.

Farmscape

The *Farmscape* cast:

Donald—failed family farmer who remembers farming with horses. Age: 68 years old.

Daphne—Owners of Montebello B&B. Age: mid-60s.

Jaime—Daphne's husband.

Randy—Owner of an independent hog confinement operation. Age: late 40s.

Kristi—Randy's wife.

Randy's parents—Mother and Father.

Lonna—Owner of a small organic vegetable truck farm. Age: mid-40s.

Joe—Lonna's husband. Age: mid-50s.

Mike—1,700 acre farmer. Age: 51.

Martin—Failed family farmer. Age: 50s.

Max—Owner of the White Oak Vineyard. Age: 70s.

Winery Owner.

Max's son-in-law.

Customer.

Nate—Monsanto employee. Age: late-forties.

Jon—Slaughtering house worker. Age: 20-something.

Martin's Mother.

Bank Official.

The following cast members may take multiple parts:

Joe: Customer, Winery Owner.

Mike: Max's son-in-law, Bank Official.

Daphne: Randy's mother, Martin's mother.

Nate: Jon.

Randy: Max.

The first production of *Farmscape* was at the Wildness Conference, February 19, 2008, in the Maintenance Shop at Iowa State University. The original playwrights and cast included:

Donald—Brian Burmeister

Daphne—Laura Sweeney

Jaime—David Perez Gomez

Randy—Daniel Wise

Kristi—Rebekah Bovenmyer

Lonna—Breget Medley

Joe—Dallas Schmitz

Mike—Brett Bender

Max's son-in-law—Brett Bender

Customer—Brett Bender

Martin—Andrew Judge

Max—Daniel Wise

Winery Owner—Jason Arbogast

Nate—Jason Arbogast

Jon—Jason Arbogast

Parent (Mother)—Laura Sweeney

Parent (Father)—David Perez Gomez

Mother—Mary Swander

Stage manager: Esther Yoder

Note: The *Farmscape* script contains some graphic and adult language.

Farmscape *is performed as a readers' theatre production. The actors read from a script. The production requires no set, props, or memorized lines. Actors sit on stools. The stage is a black box. The play begins with a projection of a farm. As bidders at an auction, the characters wander onto the stage from the audience. The actors have their bidding numbers in hand or tucked into a pocket. The actors hold up their numbers to introduce themselves and to bid on items at the auction. The characters are re-creating Martin's auction in his memory.*

(Auctioneer Sound Cue.)

MARTIN:

(Stands throughout the whole play, stage right.)

Massey Ferguson 35 loader tractor…

18-foot Kewanee disc…

13-foot DMI chisel plow…

8-foot International chisel plow…

John Deere sickle mower…

PTO driven generator, and…

my 20-foot harrow and the harrow gater…

MAX:

(Drinking wine from a glass). Now, I have to ask, are you old enough? I have a grandson who goes to the university, he's not old enough yet. He's twenty. He'll be old enough next June. I'm Max. Owner of The White Oak Vineyard. NUMBER 74.

MARTIN:

It's an auction, you know. My auction. People come in and…everything goes that day. You're surrounded by family, friends, and neighbors who are there to support you, because they understand the difficulty of the situation that you're going through…but they're also there to look for a good deal. Boy, they'd just as soon buy that tractor a thousand dollars cheaper than pay a thousand dollars more!

NATE:

I grew up on a family farm just north of here with four or five hundred acres. The farm crisis of the eighties took a toll there. I started farming back in '75 and I just didn't have enough time to build the equity I needed by the early eighties when the land went down. I couldn't compete. But, in the net end, it might have been for the better. Cause I ended up going to college then, and studying Ag., so, getting a job in the seed industry I stayed with agriculture. My specialty is producing hybrid seed corn. NUMBER 75.

MIKE:

Mike. That's my wife, Stephanie. Oh, and this is Sydney. (*Whistles.*) She's a mix between chocolate lab and something else. I farm 1,700 acres. We're at least half rented half owned. My father farmed and my grandpa did, too. I started in '79 and really had nothing to lose because I really didn't have anything going into it. Used dad's machinery. So I just grew up in it. It's all I've known. I've been through about 28 harvests, and I hate to say it but you kind of get to the point where you say, "Well, how many more am I going to be able to do?" I'm 51. And the older we get (*Looks at Stephanie*) the more you think about that. NUMBER 51.

LONNA:

I have been in my business as a farmer for about nine years. NUMBER 9. I've always been a gardener, but I've been growing food for other people for about nine years. I like to grow this stuff, a lot of lettuce, cooking greens, herbs, lots of onion, garlic, and leeks, and lots of heirloom tomatoes. I'm Lonna. This is my husband Joe.

JOE:

NUMBER 10.

DONALD:

NUMBER 68. Donald. I'm 68, almost 69 years old. I've lived in this state my whole life, and I can tell you times change.

MARTIN:

John Deere 4030—which, didn't mean a thing to me…John Deere 4020—which meant everything to me.

DONALD:

My dad and granddad farmed in the 40s. They were the last of their breed. The horse would pull these mowers behind and there would be a seat for the driver. I mean everything was basically done by hand or with horses. For years. This was late 40s, early 50s. They were the last of their breed in this area. There was simply no money in the family. They never planted corn with tractors.

MARTIN:

The 4020 tractor had a lot of symbolic value for some reason…I don't know why…I planted with the 4020…it was my original tractor when I started farming…and, there's something about the 4020… its sentimental value…it has a representation that says, this is what farming is. I mean, it was sex… there's no doubt about it…the John Deere 4020 was sexy. Yeah…it was the tractor that everybody wanted to drive, you know…you were out there in the open…it didn't matter if you got rained on or what not…

MAX:

(*Pours wine into glasses.*) Prairie rose. This one, believe it or not, we had the other night with pizza. We decided that we're not having pizza and beer any more. We're having pizza and wine. It's sweet enough that it kinda offsets the spiciness of the pizza. It's not extremely sweet, but it's kinda neat. We just have fun with our wines. People come in and say...

CUSTOMER:

Ya know, I don't know much about drinking wine.

MAX:

I say, well, you don't have to. Find one you like and DRINK IT!

DAPHNE:

I'm going to make coffee and see how many people are takers. Cream? Sugar? I'm Daphne, owner of a B&B. NUMBER 2. Ours is a concrete house because we want it to last forever! Mexican style, colors everywhere. We built it ourselves.

JAIME:

And we think that we can keep this land, not looking like a farm, but looking like a forest, or a garden. NUMBER 3.

DAPHNE:

Jaime studied for two years because we wanted light. That was part of our goal. And so he studied how the light and shadows were, so we would have all the light we could all day long. And we even positioned the house so that the east would be here, four windows worth of light. We wanted to bring the outdoors in.

KRISTI:

We both take care of the outside, whoever has time. I work off the farm at the university. I'm Kristi. NUMBER 19. I grew up in Omaha. I met Randy. I was 19.

RANDY:

We have 750 sows. We farrow every week, every day. Farrowing is having the baby pigs. And we sell every week. We sell right about 300 every week to independent farmers. NUMBER 20.

KRISTI:

His dad grew up here. He grew up here. And our kids grew up here except for about four years. They lived here their whole lives. His mom helped on the farm and she drove the tractor, but we needed health insurance. We need the pension, the benefits, but mainly the health insurance. And so that's why I went to work years ago. When I went to work we were paying almost $600 a month in health insurance. That was 20 years ago.

RANDY:

We've always had pigs here. And then, 12 years ago we went on contract. We build the building and somebody else owns the pigs, and they supply the feed, the vet, and everything. They pay us so much per pig.

JON:

So they bring the pigs through on, like, this conveyor belt, right? From the truck to the conveyor belt. Then, zap, bam, they're dead, you know? Then somebody, I don't know who, it isn't my job—somebody damn near cuts their heads off so the blood can drain. They gut them too. That's the first thing I remember from the tour. All those fuckin' heads just hangin' there. It surprised me then. I don't care though. My job is to cut parts of them up later. NUMBER 103.

JOE:

(*Clears throat.*)

JON:

Oh, you want me to introduce myself? My name is Jon. I work at Tyson—the slaughter house. I cut up pigs.

MARTIN:

Chevy pickup truck 66—red and white…
John Deere 3020…
Farm hand grinder mixer…
Case skid loader…
N Ford tractor…
Field cultivator…
Mower…
International planter…

DONALD:

I remember when I was a boy they'd pick the corn by hand. What they had were these narrow, square-box wagons—with iron or wooden wheels—that the horse would pull. You'd take the corn, throw it in the wagon and when they'd walk past the wagon they'd holler. . .

JOE:

Giddy up!

DONALD:

And the horse would go, and when they'd want it to stop they'd yell...

JOE:

Whoa!

DONALD:

Then at the end they'd scoop the corn into the corncrib. It was a lot of work in them days.

MIKE:

A lot of times I just say that I'm a kid with a 1,700 acre sandbox. I like my toys. Boy, you saw the combine sitting out there. I got big toys to play with.

DONALD:

And grandpa used to have a hell of time with the horses. They just wouldn't do what they're supposed to do. But horses were everything.

They planted corn with horse-drawn corn planters. They cut hay with the horses, too. A sickle mower. They went from using the hand scythes to using these early sickle mowers.

MIKE:

I'm just a kid with a 1,700-acre sandbox. A lot of days that's how I look at it. Just out here having fun (*Laughs.*) It's just kind of funny sometimes how these things fall together.

DONALD:

Then 15 years ago I got the harebrained idea I could be a farmer. Bought a couple of small farms. Eighty acres, farmable. We tried to operate cheap as we could. We couldn't make any money either. You just can't. If you're a little guy, you can't. By the time we paid for the machines and the seed and the fertilizer and everything else, some years we were lucky to break even. So after 10 years of trying to be a farmer, I got out. Sold the farms.

MIKE:

For starting in '79, I really had nothing to lose because I really didn't have anything going into it. I didn't have any down payments on anything purchased. And things were overpriced which led to the crisis, but you take zero away from zero and you still have zero.

DONALD:

I honestly don't know how the big guys do it either. By the time you buy the combine, the grain-head, the bean-head, you're talking half a million dollars. That's big bucks. You're talking $150,000 for a corn planter, maybe even more. Then you still need 4-wheel-drive tractors and so on. A million dollars you couldn't buy all the implements you need. And for a million dollars worth of equipment you got to farm a lot of acres.

MAX:

We talked to a number of wineries around the state who had built recently, and each of them said the same thing, they said…

WINERY OWNER:

We did not anticipate how fast we'd grow, and now we're too small and have to add on.

MAX:

So we have a goal set to be around a 20 or 25,000 gallon winery someday. We didn't start there. We started around 5,000 gallons.

NATE:

The farmscape is changing, but it's due more to business reasons. Biotech isn't the main reason, business is. The farmer is able to cover more ground with the technology, not just the biotech. The output that each acre's producing is increasing by tremendous amounts. Look at your yields.

DONALD:

As you get the bigger tractors you need more land to pay for the cost of it all. Most of these big farmers they got 3 to 10,000 acres. The whole thing's crazy. Everybody buying up everybody's else's farms. And for what?

LONNA:

We grow things here on about an acre and a half.

NATE:

What's an acre of corn producing today versus 10 years ago, versus 20 years ago? You've gone with the open-pollinated stuff with 30 bushels an acre to the state average of 180. The production's kept the price low, and that's enabled the corn to be used for things like ethanol or corn sweeteners.

LONNA:

We have a small place here. . . It's a big, big garden. And if you manage that intensively, you can grow a lot.

DONALD:

Back in the 50s if you got 60 bushels of corn to the acre you thought that was fabulous. A man could take care of his family with that. Now some of these guys act like if they don't get 200 bushels an acre they've got no crop at all. And they get it by planting the rows closer together and dumping all that fertilizer, pesticides, herbicides, what have you, all over everything.

NATE:

I mean, I'm not knowledgeable, but I understand that a lot of the corn is going to the sweeteners in your pop.

LONNA:

We don't use herbicides or pesticides. Because we're growing market garden vegetables, we're not growing corn and soybeans for the commodity market. We're growing for local restaurants, and local people.

NATE:

So far, the major commercial advantages of the Bt corn that have been brought to the market are…herbicide tolerance and insect resistance. There've been a couple major herbicides, but obviously Roundup is your big one. That's Monsanto's, the one we brought out. It kills everything except for the target crop…the one you want to save.

LONNA:

We do use compost.

NATE:

The first one that was brought out was corn-borer resistant. And now they've brought out root-worm resistance. Those are the ones that have hit the ground commercially and actually made an impact. And the herbicide tolerance…it's made an advantage in how we handle the weeds. It's cleaned up the fields and it's making it a little cheaper for the farmer to produce.

LONNA:

Compost is a good thing. We use that a lot. We don't use the synthetic stuff. We don't use the chemical stuff. There are a lot of natural pesticides that we use when we have to. Compost is like manure and chicken bedding and stuff that is mixed back in.

RANDY:

I got a lagoon up there that will hold probably 3 million gallons of liquid manure.

LONNA:

They hold manure in big liquid tanks and that's what that stuff is. The earth can't handle all of it.

RANDY:

We got tiles runnin' back there into that lagoon. We'll empty that out here in the fall, put it on this land.

LONNA:

Compost is the kind of stuff we use a lot because gardening, farming, is a renewal process, and compost is about putting back in what you take out. It's putting the organic matter, the growing things, the living things, things that the earth needs back in when you take them out.

KRISTI:

It's just a holding tank, and there's no seepage out of it or anything like that. It's real good clay that's packed in.

RANDY:

I'd say it's roughly 100 feet by 300 feet, you know, about a football field, and it's 15 feet deep.

JOE:

We shouldn't raise animals that concentrate manure that way in large lagoons. Lagoons should be outlawed.

RANDY:

Since we only have 750 sows we can put the manure on heavier. DNR will tell you how much you can put on, so much per acre. We can put all that manure on this 120 acres.

KRISTI:

Extension came out and used us and Randy in a training video for manure management.

RANDY:

And really it's nice to have this manure out of the building. There's the smell. We can smell it, too. And when they're agitating or taking this manure out of the building, if you don't have the windows or curtains open, you'll kill the hogs.

KRISTI:

Or yourself if you're in there.

JOE:

We can spread animals out, put them where their manure is handled, through combining it with organic matter and bedding. We shouldn't confine animals inside a building which is on top of a large manure pit.

RANDY:

That's when you're stirring. That's the manure underneath the building. It's much more potent and you put a lot less on a field 'cause outside the sun breaks it down, you know, rain water and everything else in there, so that's good and bad, I guess. There's a stink with any building.

KRISTI:

That's livestock.

LONNA:

If a hog is living in shit, first thing you say, it's not good, of course you're going to say it's awful. My response is to take the hog out of the shit and clean him up. The industrial model that only looks at money, has blinded itself that the hog is standing in shit. All it says is let's make the shit smell better. They're doing things at the university and other places to make the shit smell better. The downside is starting to mount up. Things will change.

JOE:

In order for the animals to remain healthy enough to be butchered, they've got to be fed antibiotics, and that's not good for the animals or the humans, because we're becoming resistant to the antibiotics.

LONNA:

I don't know why people want to deny the fact that eating meat that's been treated like that will have an effect on the eater.

JON:

My job is… well, it's damn gross. But, yeah…so the dead gutted pig body is hangin' from these clips on this line, right? And it moves from station to station.

LONNA:

How we treat creation is part of our humanity. I believe that anything we kill and use should be slaughtered respectfully, quickly, and well.

JON:

So the pig comes by in front of me. It's gutted, the head is hangin' by a string, and different parts of it have been kinda strung up different ways, all on different parts of the line.

LONNA:

I don't raise hogs myself. I do raise other animals, I do eat them, but not hogs. I don't eat hogs. I speak as a consumer, and as a citizen of the earth. It's immoral what we're doing to hogs and it's immoral what we're doing to ourselves creating food that is marginally healthy.

JON:

My job is to yank the goddamn intestines right out of the body. I pull 'em right out, but I don't detach 'em from the body. Somebody else does that.

LONNA:

I think that more people should slaughter their own food. People who eat meat should help at least once. You got to know this stuff. It's life and death.

DONALD:

I helped my grandson one day this summer. He needed help butchering chickens, and I used to work in a factory that did this. Back in the 60s. I'd do a couple hundred a day. Hell, that day I helped him we picked nine. Nine! We could have picked 50 in the fall, as long as it took us. We did it all wrong. That was terrible that day.

JON:

My job, all day long, with both arms, is just to YANK the damn intestines out of the body. The bitch of it is, too, that I'm so used to it, I don't even care anymore. I'll be thinkin' of, I dunno, *Everybody Loves Raymond*, or *The Smashing Pumpkins*, all while I'm pullin' intestines and splatterin' blood.

DONALD:

I did the killing that day. I'd hold 'em between my knees and bend their head backwards until the neck breaks. That's the way you do it. Cut their head off. Let 'em bleed out. That's a couple minutes. If the neck turns the right way, you'll get sprayed. It's a messy job. It's not a pretty thing. After they bled out, we held 'em by their feet and dipped 'em in a bucket. Hot water. Like 160 or 180 degrees. That loosens the feathers up. But ours hadn't feathered out. Couldn't get your hands on the feathers at all. It was terrible, just terrible. Factories got these feather-picking machines. We don't. So you gotta do it by hand. Nine chickens in seven hours.

LONNA:

Everything is on the edge. Everything is being pushed to the edge of safety, the edge of morality, the edge of sustainability, it's being pushed to the edge just for economic gain. As a community, we are endangering our humanity and so many good things for monetary gain. The way hogs are raised in these confinement facilities, it breaks my heart.

JON:

I have to wear this rubber raincoat deal all over me, the rubber raincoat and the goggles and the helmet. It's funny too, at mealtime, 'cause I work during breakfast and lunch and I'm just covered in blood, man, just fuckin' covered. And I eat alone everyday because nobody wants to sit next to the blood guy. Even the other guys with the blood jobs, you know? We all don't want to think we look that covered in shit. Even blood guys don't want to sit next to blood guys.

MAX:

Hog confinements? I think they're great (*Winks*) as long as they're not in my county. Actually we do have a hog confinement place just about three-fourths of a mile south of us. And yeah, we get some odor from that, especially if he's just cleaned out his facility. But most of the time we don't even notice it. Do I worry about contaminants? No, I don't know how you'd contaminate grapes. It's all just fertilizer anyway.

(*All sit. Slides and music.*)

(*Auctioneer's Sound Cue.*)

MARTIN:

Everything is selling at auction.

ALL:

AS-IS.

MARTIN:

You will be allowed 30 days for removal of equipment.

DONALD:

Be prepared to load yourself.

DAPHNE:

Hesston 1007 disc mower…

JAIME:

Like new.

RANDY:

International 966 diesel tractor with cab…

KRISTI:

Bad clutch.

LONNA:

Westendorf loader…

JOE:

Newer.

MIKE:

Vermeer Model 605 round baler…

MAX:

Vermeer 605 D baler…

NATE:

John Deere 400 grinder mixer…

JON:

Schultz Spreadmaster manure spreader…

JAIME:

We plan to build a Latino cultural institute on our land. We're going to bring arts and artists from overseas—expositions of painting, music from Latin American countries. To create a better understanding between both cultures. Because there are other Latinos besides the chicken-plucking Latinos! Latinos that are cultured, you know?

DAPHNE:

I don't know what this land was before they dug it up 150-200 years ago. But we like to keep the view as nice as we can. You can see the stars. And you can really relax. And hear the sounds of nature. And we really feel like stewards of this place.

MARTIN:

I got into farming out of pride. The concept of farming wasn't the mistake, but trying to do something because of pride…that is a mistake.

DAPHNE:

The man that sold us this land, he owns a Century Farm. He still owns the last piece of it. This is part of it. I saw the abstract yesterday.

MARTIN:

Dad had decided to file for divorce…And Mom was left alone with my two little sisters—13 and 8…and, rather than doing the smart thing of taking the money and moving on to start a new life, I was determined we could make it.

DAPHNE:

The first people on this land were the indigenous peoples. Then the U.S. government granted title in 1854 for this piece. It makes it more special to us. The farmer knew that we were going to take care of it. He knew that this was one place that we were not going to sell off for development. (*Pause.*) And we're not going to.

MAX:

If you'll grab one of those wine bottles, I want you to look at our front label. I want you to look at the tree. I want you to look at the trunk of the tree.

MARTIN:

We come from a society…a Judeo-Christian ethic…coming out of World War II that was the idea.

MAX:

You see the cross? That's part of our Christian heritage. We actually put that in there on purpose. That is not part of the original sketch, but we said, "That's what we want." We just felt that it was important to us. You know?

MARTIN:

People doing self-sacrificing, something for someone else's benefit. And I just felt that is what I was called to do. That's just what I chose to do. Well, the option at that point in time for Mom and the girls would have been to sell the property, eventually, and for Mom to become a full time, working mother, without having that family connection with the girls.

LONNA:

Growing a garden is kind of like growing a family, everything is at a different age, everything is at a different stage. It is a good thing for women to do because women are good at micro-managing, and everything is different.

DAPHNE:

We want more structures, because we have had a dream of a cultural center, a Latino cultural center, for a long time, for 20 years. And it's being lost! We don't teach our children that culture. What we would like in the end is for our land to be taken care of by the board of trustees. And our family.

MIKE:

Go back 20 or 30 years, everybody worked together. Shared machinery. Neighbors bail hay; two or three neighbors would get together, they'd bail yours, they'd bail yours, they'd bail mine. (*Pointing around the room.*)

RANDY:

I graduated in '74 and farmed ever since then. We went on contract in '89. In '95 we built these buildings up here. Before we went on contract, we were farming our land. The company said...

JOE:

You get out of the land, otherwise when spring and fall come, you'll be out in the field and not worrying about the pigs.

MIKE:

As time moved on everybody got more independent. If I had my own I could be doing mine instead of being over here working with them, working on his, while mine sits.

KRISTI:

But we don't necessarily farm the land we still own. It's rented out now and then our son will farm it after this contract is up.

RANDY:

Corn's grown on it and then I'm gonna buy the corn from the renter.

MIKE:

You just kind of lost the, not desire, but reason for everyone working together so that everyone could get along together. That's all kind of changed.

KRISTI:

Some of the bigger hog confinement companies, they just put up the buildings anywhere and people come and work it. But right around here, a place a mile to the west, they have I don't know how many buildings, but Mark's—the farmer a mile to the west—is another Century Farm, he's the third or fourth generation but he's all on contract.

JON:

It's a shit job, you know? The work is fuckin' hard, and that would be good if the pay was good. But the pay is not good. I do good, I don't have kids or anything, so I'm good, but some of these fuckers have five, six kids (*Laughs*). And that's not even countin' the Bosnians. I don't know what they do. They stick to them, I stick to me. You know?

KRISTI:

My son-in-law and daughter are another third generation, but the only way they can do it is through contract farming.

MARTIN:

I mean, I was literally insane. I would get up extremely early in the morning…I had probably 160 sows, approximately 1,200 hogs a year. I even did it for a while without a tractor to help scoop manure…I did it all by hand…loaded it into the manure spreader by hand! And Mom kept yellin' at me …

MARTIN'S MOTHER:

You're nuts! You gotta have a tractor!

MARTIN:

Everything was in corn back then—I had 60 acres of corn a year and 20 acres of oats for the straw. Land values at that time were about 3,000 an acre…and that over inflated land price within 10 years led to operational interest loan rates of 19 and 3/4 percent…I still have those loans in a box under my bed.

KRISTI:

The neighbor to the east has their third generation son starting with them, and they're big in crops, but they're also contract livestock.

MARTIN:

The reason I first left the farm was because I finally reached that point where I'd had enough...of Dad...I'd had enough broken promises and I just realized the relationship wasn't going to be what I'd wanted it to be. Sometimes it takes a long time to realize that things won't change.

JON:

My co-workers? I had suspicions with the Bosnians and the Mexicans, you know, feelin' unsafe 'n shit. But they're just makin' a livin'. I guess if I'm bein' honest, um, yeah, uh, I'd rather be by more, more, uh, workers like me. Who wouldn't?

MARTIN:

I mean, no, you can't put a value on land that's been in the family...: but that's always been the problem with farmers. Farmers have always developed a weird type of bond with an inanimate object, you know. There...there's a bonding that takes place between the land and the person that is almost the same as between a husband who loves his wife. There is. There really is. There's a strong bond there.

(*Music and Slides.*)

DONALD:

We're auctioning a horse drawn grader on steel.

DAPHNE:

Bush Hog 3 pt. mower...

JAIME:

Grain-O-Vator Series 10...

RANDY:

New Holland square baler...

KRISTI:

A.T. Ferrell & CO #27 seed cleaner...

LONNA:

5 tine forks…

JOE:

2 dump rakes…

MIKE:

A.T. Ferrell "Clipper" seed cleaner…

MAX:

3 pt. blade…

NATE:

3 pt. fork mover…

JON:

A lot of scrap iron!

ALL:

85 wrapped round bales of hay!

MIKE:

One of the foreclosures I went to, everybody agreed to not bid on anything—at a bank forced sale. If nobody's buying it, they just stop the sale…

BANK OFFICIAL:

Nobody's bidding so we're going to cancel it.

MIKE:

Had friends and neighbors looking out for him. It put the bank in a bind. At least we liked to think so. I guess we really didn't know what happened to everything after that. But everyone thought they were doing a good thing and helping out.

MARTIN:

When people know you're selling cause of divorce, they bring a different sight to the sale...it's not one of unification...but when you're selling cause of foreclosure it brings more unification. It's more of a throw back to the milk riots in the thirties of we're not gonna let the federal government take the man's estate. Because, people have the attitude of who's at fault here? It's not the fault of the person farming... it's the fault of the banks, it's the fault of the federal government, it's somebody else's fault.

JAIME

Well, we have some issues. Mainly due to progress. We were in the middle of the developer, the rural water company, and the city. So we were encouraged to sell the land at first. And that's because we were in the way. We were. We didn't want to sell the land, why should we sell the land?

MARTIN

I'm glad we still have the land—I'm glad for that, very glad for that. And that's part of the reason why I still struggle to make the payments...because, it will finally be paid for within my lifetime. And then, hopefully, somebody else who would like to become attached to it, in the family, would wanna…take it on somehow.

NATE:

The technology and the machinery enable one person to cover more ground, and that's what's changing your farmscape. If one operator can cover…the Farm Bureau has nice numbers on what a farm…a family unit is.

JAIME:

But our dream is just to maintain the land as free from any encumbrances as possible. And make it more agreeable to nature. It's beautiful in the winter, you can really go cross-country. Have foxes playing in the snow there. As soon as you go by the apartment complex it gets dark and I love it!

NATE:

There's a lot of economics to size. And so, you're seeing 5 and 6 thousand acre farms, but if you look at it it's a brother and two sons, so actually that five thousand acres is still feeding four or five families, but it's called…Joe Farms.

KRISTI:

When we first started contracting they wouldn't loan us money unless we were contracting. You had to get bigger or get out and move to town. For some people that's just not an option. It's just not in their blood to work in town.

MARTIN:

Oh, yeah, I knew there was a difference between my town friends and me. I can remember the day—I was around seven, eight, and our buddy Steve Stutsman called and asked if Ben, my older brother, and me wanted to come into town for a movie, said his mom said it was okay, so we begged Mom to drive us into town. But see, there wasn't time to shower up and we had just finished our chores, so all we had time for was a quick change of clothes, and this kid in front of us turns around right before the movie starts and says...

JOE:

Somebody stinks like hogs.

JON:

So I have a date awhile back—yeah, I had a date, fuck you—but Friday comes along and all of a sudden I can get some overtime. Oh, our date is on Friday too. So I have a choice—date or overtime, right? Well, I think "fuck it. I'll do both." So I call her and push the date back a couple of hours, do my thing here pullin' guts 'n all that, then hop in the shower and go on my date. Only thing is, I didn't get to really shower. I mean, I showered, but I didn't really shower. Normally, if I'm goin' out, I take one helluva shower. I soak in the tub for about 50 minutes, then shower for about 45, go through a whole bar of soap, the whole deal. That way I don't smell like I've been crawling around inside a dead animal all day.

On this date night I got maybe a 15-minute shower. 15 minutes! That was a bitch. The girl was nice I guess, she didn't really say nothin', but you could tell she was uncomfortable. She didn't say too much at the bar, since there was so much damn smoke. Half of it was from me, I was smokin' like a goddamn chimney, tryin' to cover up the smell. We left the bar, though, and she looked at me like I was shit in a microwave. I tried to get a little somethin' goin', but, you know, didn't work. I ended up spending most my overtime on that dinner anyways. Sucked. But, yeah. That's my shit in a microwave story.

MARTIN:

And he looked right at us and then said it again…

JOE:

Somebody stinks like hogs!

MARTIN;

Oh, we couldn't let that slide and I can remember Steve's mom coming back down the aisle with popcorn and grabbing us off him and telling us we had better get back in those seats.

MIKE:

I've always been told that everyone's a heartbeat away from losing a farm. If someone dies, the farm could get sold off. So everybody could be a future landowner that you rent from. So I try and stay in good standing with everybody, because everyone's a heartbeat away from losing a farm.

KRISTI:

At some point in there my father-in-law had expanded to bring my husband into the farm so they mortgaged the farm and I'm not sure of all the details at that point but it ended up they lost the farm and started from scratch. They were able to work out a deal that they bought it back again. They paid off all the debts they had on it and had to buy it again. They're still paying for it now 25, 30 years later. Randy will probably shoot me if I say this…All of a sudden, one day he came home from where we live now and he said…

RANDY:

Mom and Dad found a place in town, they're going to buy it. We're moving to their place in two weeks.

KRISTI:

Oh, do I have a choice? No.

RANDY:

Someone has to live there with the pigs, and Dad doesn't want to be the one in charge. Our pigs are there. We have to live there. Call the realtor. We've got to put our place up for sale.

KRISTI:

There was no discussion. This is one thing I'm still a little bitter about it, rightfully so I think, because I didn't have a choice. I was told I have to move, and it wasn't so much my husband telling me we have to move, it's his folks telling me we have to move. I think if they would've said it all different, saying…

PARENTS:

We're going to move to town. We really need you guys to move down here.

KRISTI:

That would've been totally different, but no, it was not presented that way, and I wasn't a happy camper. We planted trees the day we got married. We planted trees the days the kids were born. We just had a lot invested in that place. But it was a necessity. We had to be where the pigs were.

MIKE:

So we rented this for quite a few years. About 10 years ago we bought the acreage, built a house, and just this summer we bought the farm to go with it. And then, as the years went on, added a farm here and there. Bought the relatives out when they settled Grandpa's estate, and so I guess here I am today.

KRISTI:

The farm crisis was hard. I could probably sit and count the acreages and the farms around that quit farming in the '80s. Some of those that were just a little bit bigger back then had the capital to buy everybody else out.

MARTIN:

Coming back to the idea of farming wasn't a hard thing to do…the idea of coming back and trying to make a go of it. But I just shoulda looked at the reality of the situation. It was a pride issue because when you looked at the value of what land was, and you saw where the interest rates were, a smarter person would have taken a look at that and said, ok, cash in the hand would allow me to start, would allow mom to, possibly, start a new life and provide for the girls.

DONALD:

We make better money sitting on the value of those farms in the bank than we ever could farming them. It just makes sense—which don't make sense. Make good money for doing nothing, but work like hell and be lucky to make a cent.

KRISTI:

We just switched from everybody having diverse farms to some people getting real big in grains. Some people getting real big in livestock, and some of us doing whatever it is we can to hold on. And some just folding and saying to heck with it. So yeah, it wasn't a fun time.

MARTIN:

It's changed—the house is still here, but when I'm on the place I can walk the lane still, even though the lane isn't there. I can see myself crawling through the weeds when I was five years old and having chiggers bite my legs and going back through the ditch to explore… and Mom giving me the spanking of my life for not telling her I was going back there…and calling in the neighbors. Even though I'm not farming…I still feel a sense of attachment to it.

MAX:

Actually we're quite a ways from any communities that are expanding. We do have a number of new houses, I hope you noticed. Our vineyard has helped that. Particularly right here, across from this last field that we planted. There were people who built houses there because they knew there wouldn't be other houses there, and they liked looking out at the vineyard.

JAIME:

Somehow in the tax assessment somebody had crossed out agricultural, and just slapped residential on the entire 19 acres of land. So, by the pen, you know 'by the stroke of a pen', they changed the classification of the land. Well, see, the reason we were under the gun is because they wanted to build 33 houses, and they already built 758 units, that's a lot—and we were in the way!

ALL:

Thou shall not be in the way!

MAX:

Now with grapes you don't have crops for the first four years. So we had four years of all the money going out, and nothing coming in. So, we decided pretty quickly we weren't going to make a whole lot of money selling grapes, we need to turn it into wine, so we built a winery.

JAIME:

The pressure was for zoning. One of the arguments that some of the planning and zoning commission had against the idea of a bed and breakfast was they wanted to preserve the "pristine" environment of the rural environment. And I said, "pristine?"

MAX:

The farmers in the area know that we're here. We've contacted everybody and let them know that we have grapes. Grapes are very susceptible to 2,4-D drift. And the people who do aerial spraying, they're very careful about it. They either spray when the wind is blowing away from us, or when it's really still.

NATE:

The major alternative to Roundup was atrazine, and that lasts forever. There are studies that show it taints groundwater, so there are areas in the U.S. that are limiting atrazine.

LONNA:

In Europe it's called the Precautionary Principle. In the U.S. it's not used very much. It's about being a little more careful with what we do. The Precautionary Principle says let's stand back and err on the side of being careful. I think that we in the U.S.A. haven't done that. We're moving way too fast.

NATE:

Roundup is an excellent alternative because, environmentally, it's a better product. And from the perspective of the insecticides, the Bts—

they're the ones that were resistant to the corn borers, from what I see from what's actually in the field, using modified crops is so much more environmentally friendly.

LONNA:

Europe won't take our GMO food. A lot of people say …

ALL:

No, I don't want to eat that.

LONNA:

It's funny but nobody listens until the markets speak …

ALL:

No, I won't buy your stuff.

LONNA:

It affects me because all that stuff ends up in corn chips, corn flakes, and everything that I eat.

NATE:

We're doing a better job with environmental stewardship by using the Bt corn 'cause we're putting a lot less insecticide out in the fields, and…and that's gotta help the environment.

LONNA:

People don't always value how much effort we put in as opposed to how much money we get. I think that's a lot of farmers' problem.

The effort to grow food, especially growing things organically, it is really hard to recoup your money. We get a lot of benefits that aren't monetary, money isn't one of the biggest things we get here.

MAX:

My son-in-law, who's also a partner in this, along with my daughter and my wife, he grew up on a farm, and he said …

SON-IN-LAW:

You know, there's got to be something better than corn and beans for small plots of land.

MAX:

So we started growing grapes in 2000. So right now we have about 5,000 plants on nine acres.

JOE:

It's hard work and we have to come up with the energy to do the work. Beyond that I don't think I experience a lot of downside. It's somewhat frustrating that people who work in town at a desk don't often appreciate what goes into the food that we grow.

MAX:

You don't realize how many 5,000 is until it's pruning season, and you're out there with the hand-clippers in February, pruning. Yeah, it's a lot of work.

JON:

I'm fine with the hard work. It's just…how much do I make a year? Don't ask me that. I don't make enough to take a goddamn vacation. It's hard work, you know, uh, has-to-be-done type shit, but I can't ever afford to get away from it.

JOE:

Americans expect food to be cheap but that doesn't make sense as a producer. We don't serve poor people because poor people can't afford our food. It's not quite a downside but it's frustrating and disappointing to me that there's not a way for more people to have access to locally grown food.

JON:

My body hurts too bad to play sports—I sucked at football, you know, but I liked it—and uh…um…no, no injuries or shit. Just tired. At training, yeah, that's right, they tell you to take a shitload of Tylenol, or whatever, days before you start. You're doing the same damn thing everyday, which is crap for your body. I took hot baths every night for about a month 'till I got used to it. You know, bring a TV in the bathroom. It's not so bad. Everybody I know is graduatin' and gettin' jobs and movin'…I'm…and I'm…Yeah. I can't get away from the packing plant.

MARTIN:

I think that's the mistake that we as individuals often make…we have that myth in our head of always going West, of always overcoming, of always going to the moon, conquering space…

DONALD:

Farming business is just like any other business. It evolves where the profit is.

LONNA:

It's wonderful because every plant has its own little thing whether it's lots of rain, cool feet, cool things around their roots. Whether they like hot, like peppers like hot. Everything is different.

MARTIN:

We have that same idea that takes place with trying to farm, that we can always overcome, that we can always roll the dice, and produce a crop that will end all our woes.

LONNA:

A tricky thing to grow is lettuce, lettuce greens, because they like cool. They're delicate. They're like a spring day.

MARTIN:

I mean, I can remember praying for 80-bushel beans! It just doesn't happen.

LONNA:

And everything starts out beautiful here in the spring. But, along about June we begin to get summer and the hot winds blow and it stops raining.

MARTIN:

Sometimes it takes a long time to…to realize that things won't change.

LONNA:

That is the trickiest thing for me to grow, to keep lettuce greens growing. That is the hardest thing.

(*Music. Slides. Lights fade out. Curtain Call.*)

Commentary on Farmscape— The Changing Rural Environment

How Farmscape Happened

Mary Swander

In the fall of 2007, I didn't know what was going to happen. I walked into my English 557: Writing About Environmental Issues class on the first day of the seventeen-week semester with an idea to write a play based on interviews, a collaborative project that would be staged as a readers' theatre drama. The actors would read from scripts with minimal blocking, no lines to be memorized, and simple costumes and props. In Iowa State University's MFA Program in Creative Writing and Environment, our graduate students come to Ames, Iowa, to write creatively about nature, culture, and place. This was an experimental course.

"I'll be honest with you," I told the students. "I've never taught this course before—nor has anyone on our campus."

A look of fear flashed back at me from the class. The room was filled with a mixture of ten graduate and high-level undergraduate students who ranged from 20 to 45 years of age. It was a diverse bunch with a balance of males and females, minorities, and representatives from various majors across campus.

I tried to counter my students' consternation with my enthusiasm. We were going to write a play together—professor and students on equal ground. The drama would be about an environmental issue and would be based on interviews of real people in a real conflict. The classroom would be a collaborative laboratory where we would discover new possibilities.

"Do we have to actually act in this play?" the students asked, terror once again filling the eyes of these shy, Lake Wobegon introverts.

I reassured the class that I would go to the theatre department to secure actors for the one performance that we planned to have at the beginning of the spring semester. All we had to do was write the play.

"And we have only one limitation," I told the students. "We don't have any money."

We had to stick close to home, find a local issue, one that could be probed from different angles and perspectives. As much as we would have liked to travel to the Arctic to write about the effects of the melting ice caps, or to Brazil to document the disappearance of the rain forest, we simply didn't have the money for those projects. In preparation for our work, we read *The Laramie Project* by Moises Kaufman; *Twilight: Los Angeles, 1992* by Anna Deavere Smith, and *The Exonerated* by Jessica Blank and Erik Jensen. These examples of verbatim theatre are plays about real events, all told through non-fictional research about major American happenings: the murder of Matthew Shepard in Wyoming, the police beating of Rodney King, and the sentencing of innocent U.S. citizens to death row. We used these plays as our models, analyzing, studying, and borrowing their techniques.

Gradually, we accumulated a list of possible topics on the blackboard that looked something like this:

    1. The death of the family farm.
    2. The pollution of the rivers and streams.

3. The use of genetic modified organisms (GMOs).

4. Urban sprawl.

5. Concentrated Animal Feeding Operations (CAFOs).

6. Monocropping.

7. The depletion of the soil.

8. The use of pesticides and herbicides.

9. The disappearance of small towns.

10. The rise of meat-packing plants and an influx of immigrants.

The students saw the pattern before I did.

"All of these issues fall under the rubric of the changing farmscape," a student offered.

After trying on an array of metaphorical titles, we finally settled on the more straight forward title of *Farmscape: The Changing Rural Environment.* Then we spent time learning the art of interviewing, generating common questions for our interviewees, and practicing mock interviews on each other. Once confident in our roles, we set off as a class to the Montebello Bed and Breakfast and practiced a group interview with Jaime and Daphne Reyes whose business was near campus. They graciously let us set up a video camera and recorders in their living room, serving us all coffee and tea—gestures that eventually made their way into our play.

The next week the students fanned out across the state, hanging out in cafés where farmers gathered for breakfast, driving down gravel roads in the countryside to find farmsteads, phoning relatives and friends for contacts. We tried to find one interviewee to speak to each of the environmental issues we'd generated. So, while one of the students tasted wine with the new vintner Max, another donned protective clothing to enter Randy and Kristi's hog confinement operation.

Back in the classroom, we posted the interviews on a class website and shaped each of them into a dramatic monologue, deleting the questioning, repetition, and extraneous material that had appeared in the original transcripts. Next, we caught our breath, took a step back and took a hard look at what we had gathered. Mostly, we had accumulated rich material from conventional agriculture, exposition with some duplication but without many dramatic incidents. And very little tension. Everything in the farmscape seemed just hunky dory. For example, one student interviewed his father, the manager of a meatpacking plant. The manager viewed the plant as a clean, well-lighted place with few problems. I was happy for the manager, but the interview did not make for riveting drama.

"Let's set this piece aside for now," I told the student. "Go back and interview one of the workers at the plant and see what happens."

Enter the character of Jon, a man who spent his days ripping the intestines out of hogs.

Quickly, I realized that not one of my students had grown up on a farm, although most of them were only a generation or two removed from it. And several of the students were urbanites—one from Detroit and another from Los Angeles. The class needed more background about the history of agriculture. For example, many of our interviewees spoke of the scars they had experienced during the Farm Crisis of the 1980s. I'd remembered that period vividly with its farm foreclosures, suicide hotlines, and the shooting of a banker in a small town near where I'd lived.

But my students only had a vague idea of the time. We halted work on the play for a couple of weeks and I drew up a reading list that included everything from Osha Gray Davidson's *Broken Heartland* and Michael Pollan's *Omnivore's Dilemma* to Vandana Shiva's *Stolen Harvest*. We read and discussed the books, attempting to gain a wider view of the scope of our project.

We were subtitling our play "The Changing Rural Environment." What was the farmscape changing from? And what was it changing to? What was driving the change? How were farmers and rural people dealing with these changes? How did these changes affect their families and communities? What were the wider, global implications of what we were documenting in Iowa? And why would the larger, urbanized world care?

While we read, I sent the students back out to gather more information and address some of our problems—first the lack of dramatic detail.

I instructed them to, "Go back to your people and ask them to tell a story about their farmscape experiences, a once-upon-a-time defining scene."

In came Donald's memories of butchering chickens, Martin's memory of going to the movies in town smelling of pig manure, Kristi's move to her in-laws' farm, and Jon's shit-in-a-microwave-story.

Next, we lacked female voices. Mostly, the students had interviewed men. Even when a student had set out to interview a woman, her husband often sat in on the session and did most of the talking.

And we didn't have a single farmer who was truly from the "alternative agriculture" sphere.

As it turned out, an African-American student hadn't done an interview at all. She was fearful of venturing into the Iowa countryside and knocking on a farmer's door. I arranged for her to interview Lonna and Joe, friends of the Creative Writing program who had often invited our students to their small organic farm to plant garlic and attend pizza suppers. I'd known Lonna as a rather quiet person, Joe the more vocal of the couple.

"And try to get Lonna to open up," I told the student as I sent her off down the narrow lane to Onion Creek Farm. "We need more women in the play."

In came Lonna's lines about CAFOs, GMOs, and the Precautionary Principle.

The first week of November found us in pre-writing mode. We settled on modeling the structure of our drama on *The Exonerated*, a play that orchestrated the speeches of a cluster of diverse strangers around a common experience. In class, we sifted out some of the threads of *Farmscape:* the acquisition of land and the struggle to hold onto it, the attempt to make a living, weathering the Farm Crisis, raising and slaughtering meat, and all the hard work that went into the rural life. I sent the students to the computer lab to work in groups, asking them to begin to cut and paste the various speeches together.

I circulated among the groups and the flash of fear once again met my eyes. I knew what they were thinking. This is just too hard to do. We're never going to make anything of this. The whole class is going to be a complete failure and we're going to look like idiots when and if this play ever gets on its feet. I couldn't say I blamed the students. Some of these same doubts drifted through my mind, but as a seasoned writer, I recognized this phase of the process.

"We need a frame," I suggested to the students. "Something that creates a context for the play, some dramatic action that compels the piece forward."

Martin's auction came to mind. He had been so traumatized by the loss of his farming operation that 20 years later he could still recite a list of items that were sold.

Massey Ferguson 35 loader tractor...

18-foot Kewanee disc...

13-foot DMI chisel plow...

And so we pretended to recreate Martin's auction, an event that might draw together all our characters in one place for one day. The only problem: few of the students had ever been to a real farm auction. So, I sent them off again to experience the real thing and they returned with their eyes open and real sale bills in hand.

Everything is selling at auction.

As is.

You will be allowed 30 days for removal of equipment.

Be prepared to load yourself.

Finally, I spent most of Thanksgiving week drafting the play, taking the basic outlines that the students had generated and blending them into a real drama. I wove the speeches back and forward, tightening them, cutting repetition, eliminating digressions. I found keywords that could be tossed in the air like balls, moving from character to character. I highlighted monologues. I polished, re-drafted, cut whole sections, re-drafted again. I sat at my desk stopping just a couple of hours for Thanksgiving dinner.

When we met again the last week of November, I tripped down the steps from my office to the classroom. "We have a play! " I said. "We have a play!"

That night, we combed through the whole script line-by-line, word-by-word, polishing, correcting spelling, tightening again. By the end of the evening, the students' faces and moods had brightened. Our long project, all of our footwork, research, and hard efforts were beginning to seem rewarding.

"We're going to get to take the parts in the play, aren't we?" The students asked.

Oh, no, I thought, but eventually agreed. We spent the final two weeks of the course learning basic acting techniques and rehearsing the play in a readers' theatre format. Slowly, my shy Lake Wobegonians began to venture out on thin thespian ice. Most had never been on the stage before, so we had to start with breathing exercises and learning to project our voices. We had more male characters than male actors, so we had to double up on minor parts. We had one actor so gripped by stage fright that she didn't come back to class after break. *Dear God,* I prayed one night, *please let me get through this one performance of* Farmscape, *then I'll never bother you again.*

And my prayers were answered. February rolled around and we had a dress rehearsal in a local coffee shop with a surprising number in the audience. One of the students in the class was an excellent photographer. He visited the Ames Historical Society and found old timey farm photographs from the turn of the twentieth century. He also took photos of our current landscape and others of many of the real interviewees. We added slides to our production. Other students made programs, found props, learned to run the lights and projector, and put out the publicity.

The day of our first real performance arrived. Several students ushered. The audience filled the seats at the Maintenance Shop, a small studio theatre, in the Memorial Union on the ISU campus. We were part of the Creative Writing Program's annual Wildness Symposium. The students had invited the real interviewees and many of them were in attendance. But the student with the crippling stage fright failed to show, so I took her part. Another student warmed up the crowd with guitar music and played through the opening slide show. Then we were on. And were we on. My Lake Wobegonians rolled through the performance with verve and confidence, not missing a line, playing their characters flawlessly, commanding all their knowledge of the past semester, and bringing a real depth of understanding to the stage.

The lights faded. The lights rose again. Applause. More applause. The students and I stayed on the stage. The real interviewees joined us. We had a dialogue with the audience. The first *Farmscape* "talk-back" was born.

Thank you, Lord. Thank you, I said to myself, crashing into bed that night, relieved that *Farmscape* was a success—completed and over. On to my other classes. On to my other duties. I sat in my office the next morning grading papers when the phone rang. It was Fred Kirschenmann from the ISU Leopold Center for Sustainable Agriculture. One of his staff had seen the show the day before.

"I want you to meet me for lunch," Fred said. "And let's figure out how we can get this show on the road. You can't just have one performance of this play."

A few weeks later the Leopold Center gave me a grant to tour the production to three locations around the state. I sent out an e-mail to contacts and the three locations were booked in a day. By the end of the week, five more venues had signed on. And so I began to tour *Farmscape*, piling scripts, my computer and projector, and some Amish-made fold-up benches, I used for a set, in the back of my car. I crammed my ISU classes to the beginning of the work week, put other classes online, then took off travelling the State of Iowa, from small towns to larger cities, from coffee shops to colleges and universities.

No venue was too small. I knew this show would reach its audience in a grassroots way. Many of the original students helped me out with local productions near Ames. Laura Sweeney and Jason Arbogast were real troopers, appearing in numerous productions. Claudia Prado-Meza, a student in the Sustainable Agriculture Program, jumped in and helped run the projector for several performances. She provided insight into the role of immigrants in the state—how and why they found work in the slaughtering houses. Leigh Adcock, Executive Director of the Women, Food, and Agriculture Network, helped sponsor one of the shows and appeared in the audience for the talk-back. She provided insight into the role of women in the rural environment.

The show kept touring. We pasted together a troupe. We played conferences and festivals, more coffee houses, community centers, museums, and bed and breakfasts. We even played a beauty parlor once.

More often, the venues created their own casts. A producer—or "whirling dervish" as I called them—worked real magic in every location. The dervishes gathered together willing volunteers, secured a performing space, found sponsors and funding, put out the publicity, arranged for panel discussions and displays of local foods after the show. I worked with the dervishes for months before a production, creating a checklist of things we needed to do in advance. They plowed ahead tirelessly, adding their own energy and twists on the performances. In Coon Rapids, Iowa, I found whirling dervish Karan Founds-Benton, a theatre artist from Los Angeles. She just happened to be living in Coon Rapids for a year while she worked on a novel. Soon, I began sending Founds-Benton out to direct the show. I appeared on the day of the performance to give a workshop and conduct the talk-back discussion after the performance.

I learned to be flexible. I had given one small-town opera house some of the Leopold Center grant money for a show just when the 2008 flood washed away their matching funds. I moved the money on to a nearby town that had remained high and dry. Dates jumped around on the calendar. Conflicts arose with religious holidays, sporting events, and weddings. Often I'd get a call from a whirling dervish that would go something like, "We're going to have to move the *Farmscape* show to the next week-end in July. We forgot that third week-end is the Rocky Mountain Oyster Festival."

Yet, the tour kept growing. We received some great publicity in places as diverse as *The Des Moines Register, Edible Iowa River Valley,* and *Farm Journal.* The Area Arts Council in Grinnell, Iowa, booked the show in their beautiful refurbished downtown theatre with 350 in attendance. Tom Lacina, the Grinnell whirling dervish, cast the play well with Harley McIlrath, the manager of the campus bookstore, playing Randy, a man close to his own skin. Lacina brought in the Burlington Street Bluegrass Band to play, raising the production to a whole new level. Someone saw the Grinnell *Farmscape* performance, then booked it in another location. And so it went.

Soon each venue began supplying its own music. A local band became the best draw. Some bands fit their repertoire to the show, but others, like one in Dubuque, wrote whole new tunes just for *Farmscape*. When we couldn't find a band, I sat in with my banjo. Many *Farmscape* performances were so good that the venues were asked to tour their productions. For example, the Dordt College show in Sioux Center, Iowa, toured down to the Lewis and Clark Interpretive Center in Sioux City. The Hearst Center for the Arts show in Cedar Falls, Iowa, toured down to an agricultural journalist conference in Des Moines.

Vicki Simpson at the Hearst Center for the Arts in Cedar Falls, produced the first "celebrity cast," to great results. Jim O'Loughlin, a well-known professor at the University of Northern Iowa, took on the double roles of Nate and Mike, and pulled in his own set of fans. Other venues took up the celebrity casting idea with city mayors, university deans, or well-known farmers in roles in the play. Smaller towns and cities like Independence, Iowa, used local ministers, teachers, and well-known farmers in their productions. Mary Klotzbach's essay nicely lays out the nuts and bolts of producing a town production of *Farmscape*, from finding local celebrities to booking a space.

Some venues, like Independence, worried about the use of some of the graphic language in the play. Jason Arbogast created the original role of Jon, the meatpacking plant worker. Arbogast understood Jon's character very well and knew that his rough language was part of his milieu. But for those hesitant to offend, I created a "clean" version of the script that was available upon request. The original version published in this book may be raw to some, but it retains the original power of the speech of the interviewees.

No matter which script was used, *Farmscape* made an impact—on the cast, the audience, and the wider community. Even though I usually mounted a show in less than a week, each cast bonded in a spirit of mutual professionalism and respect. They always expressed a sense of gratefulness for having the chance to take part in the production and to get to know their fellow actors. Even in small towns of less than 1,000 people, the actors said that they didn't really know each other until they had had the *Farmscape* experience.

The talk-backs after the shows allowed the audience to interact with not only the actors, but often the playwrights, interviewees, and panels of various experts. These discussions were both reflective and probing and often as dramatic as the play itself. The talk-backs went to the real heart of *Farmscape*. The rural environment is changing quickly all around us in Iowa, the U.S., and the world.

Farmscape raises key questions about these changes and about the stewardship of the land. What are the lingering effects of the Farm Crisis? What has happened to the family farm? How do we now raise our crops? How do we raise and slaughter our livestock? What are the economic and class issues involved in *Farmscape?* How are minorities and women involved in the scheme of things? Who is making it financially? How are we using our natural resources?

Sometimes the talk-back discussions became heated. Several large producers who farm 1,000-10,000 acres thought *Farmscape* was too negative and expressed their gratitude to conventional farming methods. They saw few downsides to their profitable operations.

"Farming has been really good to me," one producer said.

Several times an audience member flatly pronounced the family farm "dead."

"Small farms aren't viable anymore," one woman said. "They're a thing of the past."

This kind of comment provoked outrage from young beginning farmers who were trying alternative methods—CSAs, direct marketing, or the rotational grazing of dairy cattle as Francis Thicke describes in his essay.

CAFOs became an issue, and once I refereed a shouting match between a grain-fed and a grass-fed beef farmer.

Sometimes the talk-back discussions became heartbreaking. On a wintery night at the community college in Clinton, Iowa, a banker rose to his feet. He thanked me for coming and thanked the cast for performing the play.

Then he burst into tears.

"You see," he said. "That auction that you saw on the stage was my auction. That was my farm. Martin is my brother."

The banker was the "Ben" character mentioned in the play, the teenage boy who, with his brother, Martin, had tried to save his family farm. The audience and cast were stunned and moved by the revelation. We spent the rest of the evening discussing the ways in which each of us—whether we were conscious of it or not—had a connection to farming and the land.

As several of the essayists in this book point out, often these *Farmscape* discussions continued into the theatre lobbies, into the town coffee shops, college classrooms, and sale barns of the region. Discussions of farming led to thoughts on conservation and issues of resource management. *Farmscape* is a microcosm of rural life. Its resonances are widespread.

At the talk-back after a show in Colorado, a woman addressed cast member Ben Rainbolt, Executive Director of the Rocky Mountain Farmers' Union.

"This play presents the farming issues of Iowa, but what are the concerns in Colorado?"

"The geography and conditions change, but the issues are the same," Rainbolt replied.

Wherever the location, the way we're farming in the Midwest is part of a larger structure of agriculture in the country—a larger web of food relationships in the world. Farmers and rural citizens are not and never were isolated members of society. Anna Lappé discusses the broader implications of *Farmscape*—and why they matter. In her essay in this book, Lappé details the ramifications for the soil, water, and air, in relationships to the problem of indebtedness, mono-cropping, pesticide and herbicide use, livestock waste, and GMOs.

Yet, Lappé also sees the recent positive gains that have been made in sustainable agriculture. She concludes her piece with an image of both her family in New York and a family she visited in Missouri— worlds apart in backgrounds and perspectives— getting involved in a personal relationship with healthy food.

With healthy food growing all around me on my small acreage, I sat in my writing studio on a sunny summer morning. I'd just booked another *Farmscape* show and had arranged for Founds-Benton to take over the heavy lifting. I congratulated myself on my management skills, and turned my attention to another project, when I received an e-mail from Fred Kirschenmann.

"What do you think about starting a group that would integrate the arts and agriculture?" He asked.

Kirschenmann thought that *Farmscape's* success was a good starting point, that it could encourage a whole outpouring of artistic and agricultural expression.

"I'll open my home for a dinner if you'll invite some people for a get-together," he added.

I sent out about six e-mails, hoping to get a dozen people to attend. The night of the dinner arrived, and Fred's spouse wondered how many people to expect.

"Twenty, at the most," I told her.

Forty-five people poured through the door: farmers and gardeners, chefs and caterers, writers, musicians, theatre, movement and visual artists. The energy level was high and we soon found ourselves getting together for films, speakers, and panels. Quickly, we took up Gene Logsdon's book *The Mother of All Arts: Agrarianism and the Creative Impulse*, and used the text to generate a common philosophy. The wisdom behind this voice allowed us to contemplate our actions more deeply and embrace a history of arts and agricultural collaboration.

We began throwing Local Wonders Dinners. Jaime and Daphne offered their lovely bed and breakfast for space. We now come together a couple of times a year for a potluck dinner. We suggest a $20 donation at the door. Gainfully employed adults often offer more and struggling students less. Everyone brings specialties—from tamales to sushi—and we have a prize for the most artistic presentation. We might have a reading or a short talk about an AgArts activity, then we get down to business—the presentation of the proposals.

Any member can present a short proposal and pitch a project for the donation money. Rules: 5 minute pitch and the project needs to enhance the local cultureshed. By a show of hands, we vote on the best proposal and, right then and there, the winner takes the cash in the donation basket. Accountability: the winner returns to the next dinner and becomes our entertainment, reciting the poetry, singing the songs, showing the slides, or work that had originally been proposed. We encourage the winners to eventually exhibit or present their finished work for the general public in a local gallery or performing space.

The AgArts and Local Wonders Dinners are catching on and gaining attention. I've been asked to give talks at rural arts, agricultural, and environmental conferences about the concepts and have encouraged others to launch their own AgArts groups and dinners. The grants eliminate the usual bureaucracy of proposals, matching funds, judging panels, committees, transportation to meetings, and difficulties and delays in receiving checks. Members have "grown" their grants by applying to other funding agencies including the Iowa Arts Council. Information on AgArts and the Local Wonders Dinners can be found on our website: agarts.eserver.org

In the fall of 2007, I didn't know where the class was going to go. Now the original playwright students—those hard working, cooperative people who wrote so beautifully together—have completed their theses, gone on to graduate schools, gotten jobs, and gone out into the world with greater curiosity and confidence. They have kept in touch and helped move this project forward. I couldn't be more proud of them. We formed our own bond, and they deserve much recognition. I continue to work with many more whirling dervishes on various projects. What a privilege. The Iowa State University Leopold Center has been steadfast in its support. The College of Liberal Arts and Sciences, the English department, and the Program in Sustainable Agriculture have all helped this project find its wings. The original interviewees continue to be generous and some of them even occasionally jump in and take their own parts in productions. Many, many thanks to all.

Now, the play is being picked up and performed in venues throughout the country. It is being read in literature, writing, agronomy, and environmental classes in colleges and universities. I have been asked to give workshops on the *Farmscape* concept alone, as many locales want to write their own dramas based on their own issues. I've lectured on the concept in such diverse places as Coon Rapids, Iowa; Pittsburgh, Pennsylvania; and Ballytobin, Ireland.

Farmscape has come a long way from the days when I drove around with the set stuffed in the back of my car. And I've travelled a distance in my understanding of not only the complexities and intricacies of agriculture, but the potential of agrarian lives and communities. In these last years, I've realized that the *Farmscape* characters—from Martin and Donald to Kristi and Lonna—are there in the foreground, confronting key "environmental issues."

While many of the rest of us sit back safely in our houses or cubicles, the *Farmscape* characters are caught up in real dilemmas. These characters are dealing with quickly advancing changes in the landscape. We can learn from their struggles and resilience. These accompanying essays open the door to an even wider perspective on the issues that reverberate through this play. These pieces create yet another community, a chorus of thoughtful voices that help us reach a deeper reading of the ground under our feet.

Interested in a *Farmscape* production? Contact me through my website: www.maryswander.com and join us on Facebook: www.facebook.com/FarmscapeTheChangingRuralEnvironment

The American Farmscape and Why It Matters

Anna Lappé

In the early 2000s, I was visiting farmers in Missouri for a book I was writing on food—about the choices we make as individuals and as a society about what we eat, how we grow it, and the impact those choices have on all of us.

And so I found myself, one sunny late fall afternoon, sitting in the farmhouse of a soybean and corn row-crop farmer. As the sun traveled across the sky, he talked about the shock of volatile commodity prices, about consolidation in the seed and chemical industries, about the heartbreak of seeing nearby families lose their farms…and about how it was becoming increasingly difficult for him to put food on his table.

A farmer finding it hard to feed himself? But weren't we surrounded by tens of thousands of farm acres? Wasn't there ample food? Yes, and no.

Like most other American farms, his wasn't growing food, exactly. His farm was producing an industrial product. The corn all around us was destined for nearby ethanol plants or livestock factory farms.

Today, more than half of all U.S. corn is grown for fuel, much of the rest for factory farms and nearly all soy meal is used as animal feed.[1]

A farmer unable to feed himself is not the only paradox of today's farmscape. It's been more than half-a-century since we revolutionized how we grow food; the changes to our landscapes, communities, and farm families have been no less extreme.

The Industrial Farmscape

The revolution in food and farming has meant the industrialization of our food chain, with productivity in the fields ever-more dependent on energy-intensive irrigation, human-made fertilizer, and an often highly toxic chemical arsenal to fight weeds and pests. Industrialization has also meant a concentration in power: most seed companies are today owned by many of the same firms selling farmers a concoction of chemicals to apply to their fields and together are a $32 billion dollar industry.[2]

Remember Donald, the farmer in *Farmscape*, who laments million-dollar equipment purchases? Industrial farming often leads to deep indebtedness as farmers must use expensive machinery to tend thousands of monocropped acres. Industrialization has also meant ripping livestock from their natural habitat and diets, and consolidating them by the tens of thousands into inhumane and polluting factory farms.

This radical shift in practices that work against nature and keep farmers dependent on corporations has played itself out on the landscapes of our nation.

Some people may think of water as a renewable resource; it falls from the sky. But the water used to irrigate the fields across nearly one-quarter of the nation's farmland is drawn from a deep underwater reserve known as the Ogallala aquifer. Unlike lakes and streams, this water isn't readily replenished. Think of it as fossil water. Thanks largely to industrial agriculture, we're quickly draining it. Today, more than a quarter of the Ogallala aquifer is gone in parts of Texas, Oklahoma, and Kansas.[3] Wells have been drying up across the cornbelt and overall the water level in the aquifer has dropped by 100 feet since 1980.[4] If we don't limit how much we are drawing down the aquifer, it could be empty within several decades.[5]

Industrial agriculture not only uses up vast tracts of water, it pollutes waterways, too. Agricultural runoff, containing synthetic fertilizers not taken up by the crops, for instance, creates aquatic dead zones where algae blooms and sucks up all the oxygen, killing or driving off other aquatic life. In the United States, there are now significant "dead zones" in the Chesapeake Bay, along the West coast, and in the Gulf of Mexico where runoff spreads in the deep waters across an area more than four times as large as the BP Deepwater Horizon oil spill.[6]

Waste from livestock factory farms is also a serious source of water pollution. Waste held in cesspits on these operations leaches into the local groundwater spreading a "cocktail of nitrates, phosphorous, hydrogen sulfide, bacteria, and other substances like antibiotic drugs," says Dan Imhoff, editor of *CAFO: The Tragedy of Industrial Animal Factories.*

This problem of livestock waste is yet another example of how industrial agriculture undermines natural systems. Whereas manure on sustainable farms can be cycled back naturally into the soil as a source of fertility, in a factory farm waste has become an expensive and environmentally destructive problem.[7] Consider that just one factory dairy farm crammed with 5,000 cows can produce enough waste to fill cesspits that, if laid side-by-side, would be as large as 14 football fields.

Landscapes

Industrial agricultural methods also make our farmscapes vulnerable to soil erosion, primarily because the monoculture row crop rotation year after year—either corn-to-corn or corn-to-soybean—leaves the ground bare in most places from November to June. Industrial agriculture practices also compacts soil, making it ever more vulnerable to erosion. New data is showing that we're losing soil even faster than had previously been assumed. In Iowa, for example, a May 2007 storm caused more soil loss in one day than experts had estimated for the entire year.[8] Keep in mind that soil erosion doesn't just mean the loss of valuable topsoil, soil erosion also means much of what is sprayed on the land travels into other landscapes—into our waterways and oceans, for instance—contributing toxic runoff that has environmentalists increasingly worried.

On the other hand, farmers focused on sustainability plant cover crops, keeping their soils covered year round, to protect against erosion. Sustainable farmers also value crop diversity and nitrogen-fixing legumes to help build soil matter and increase fertility naturally.

GMOs and the Farmscape

One of the latest intrusions on industrial farms are genetically modi-fieid crops (GMOs). GMO corn and soy varieties were introduced commercially in 1996, and today, 88 percent of corn and 94 percent of soybean acreage are now planted with genetically modified varieties.[9] Proponents argue that these crops reduce the environmental impact of industrial agriculture—limiting pesticide use, for example—but the facts suggest otherwise.

Roundup Ready GMOs were engineered to be resistant to the herbicide, Roundup. But after years of use, farmers are now spraying ever more herbicides as weeds become resistant to Roundup. Indeed, "resistance to Roundup is an increasingly crippling problem for this industrial model of production," says Craig Cox, Senior Vice President for Agriculture and Natural Resources for the Environmental Working Group. Many Midwest farmers are finding that they're re-verting back to toxic chemical cocktails to address "superweeds," some unruly enough to "stop a combine in its tracks."[10] Others are aban-doning their farms all together as superweeds keep millions of acres from being harvested at all.[11] Chemical and biotech giants, Monsanto and DuPont, among others, are now patenting a new line of GMOs resistant to 2,4-D, a decades-old pesticide with a history of environ-mental contamination. "Old chemicals that we were supposed to be getting rid of because of the benefits of glyphosate-resistant crops, like Roundup, they're coming back," says Cox.

The New American Farmscape

In the past several decades, the American farmscape is changing once again as farmers and citizens are waking up to the true costs of our industrial food system—including this new generation of genetically engineered seeds—and turning toward sustainable methods. Sustainable farmers work to avoid unnecessary use of external inputs, especially those that have adverse impacts on human and animal health, and embrace instead agroecological practices to manage pests and weeds and achieve soil fertility.

Through new links between consumers and farmers, including farmers' markets and community-supported agriculture (CSA) farms, eaters are connecting with sustainable farmers. There is a growing movement voicing a desire for public policies to support these ecological methods, methods that protect biodiversity and grow food with nature not against it.

Like thousands of other families in New York City, during the harvest season, we get a weekly share from our CSA farm. As shareholders of the farm, my family invested at the beginning of the growing season—along with 222 other families—and we all benefit. My two-year-old daughter tasted her first raspberries, pears, blackberries, green beans, basil, plums, peaches, summer squash, and more, thanks to the Green Thumb Farm. I watched her eat up the food in peace. I didn't have to worry about hidden pesticide residues or the unseen environmental impact of a far-off farmscape.

It's not just urban consumers and peri-urban farmers who are embracing this shift in consciousness. The day I visited with that farmer in Missouri, he told me his children came home from college the year before and encouraged him and his wife to grow real food alongside their commodity acreage. And so, the year before, the family planted their first kitchen garden. Egged on by his children, he was putting homegrown food on the family's plates.

As if to illustrate the message, his daughter drove up just as we were saying goodbye on their back porch. She dashed into the garden and bounded up to us. In her hands she held out two watermelons—the garden's first two. With big smiles, they insisted I take one with me.

Notes

1 Lott, Melissa. "The U.S. Now Uses More Corn For Fuel Than For Feed," *Scientific American*. October 7, 2011.

2 Agricultural Chemical Manufacturing Industry. *Hoovers*. Accessed on 4/3/12: http://www.hoovers.com/industry/agricultural-chemical-manufacturing/1086-1.html

3 Pearce, Fred. *When The Rivers Run Dry*. Beacon Press. 2006. (23)

4 National Research Council (U.S.). Committee on Twenty-First Century Systems Agriculture. *Toward Sustainable Agricultural Systems in the 21st Century* (Washington, D.C.: National Academies Press). (61)

5 Ibid.

6 http://articles.sfgate.com/2010-04-27/news/20871818_1_oil-spill-shoreline-impacts-rig,&http://www.bloomberg.com/apps/news?pid=20601124&sid=a1WsUp_sIqa4

7 http://www.ucsusa.org/assets/documents/food_and_agriculture/cafo_issue-briefing-low-res.pdf

8 Cox, Craig, Andrew Hug, and Nils Bruzelius. *Losing Ground*. Environmental Working Group, April 2011. http://static.ewg.org/reports/2010/losingground/pdf/losingground_report.pdf

9 http://www.ers.usda.gov/Data/BiotechCrops

10 Laskawy, Tom. "The Chemical Treadmill Breaks Down and the Superweeds Did It." *Grist*. October 8, 2009.

11 "Roundup's Potency Slips, Foils Farmers." *St. Louis Today*. July 25, 2010.

Producing Farmscape in Independence, Iowa

Mary Klotzbach

As a hometown friend of Mary Swander in southwestern Iowa, I looked forward to and thoroughly enjoyed the presentation of *Farmscape* at the Hearst Center in Cedar Falls in March of 2009. I was inspired to invite her to Independence for a production of the play. In 2010, when she accepted, I enlisted our Local Arts Comprehensive Educational Strategies (LACES) group to help with the planning.

Laying the groundwork

LACES was established by Stan Slessor, a former superintendent of schools. The mission of this group is "promoting the arts to enrich life." It has hosted and helped fund such artists as P. Buckley Moss, Dan Knight, the Pushkin Ballet, Simon Estes, the Missoula Children's Theater, Eulenspiegel Puppets, a Waterloo/Cedar Falls Symphony summer concert featuring Independence native Chester Schmitz (formerly of the Boston Pops), a Red Cedar Chamber concert, and many others. Funding has come from the Iowa Arts Council, a local endowment, and fundraisers.

When we received the "*Farmscape* Touring Facts and Needs" sheet from Ms. Swander, we knew we had ourselves a challenge, but one worth overcoming. Like many of our productions, the key elements were talented performers, an appropriate venue, and adequate funding.

Harnessing our assets

Our first task was finding a venue for the *Farmscape* performance. Independence, Iowa, is a relatively small rural community with a population hovering around 6,000. It is the Southwest Gateway to Northeast Iowa. For some it is a bedroom community or community of compromise, conveniently located within an easy commute to the better jobs in the bigger cities of Waterloo and Cedar Rapids. However, Independence is fortunate to be located at the crossroads of two major highways (US 20 and Iowa 150), have an active rail line, and a small but constantly improving public airport.

Independence locally employs more than a thousand citizens in our light manufacturing firms, creamery, state mental health institute, downtown shops and businesses, county court house, and school system. The community supports a summer Farmers' Market at the historic Wapsipinicon Mill site with produce from local growers. Independence has also invested in a Farm-to-School program where many of the participants have shucked corn or pitted strawberries for school lunches. Children have garden plots which they tend and harvest on school ground for their own use.

Many small towns might look to their schools for a *Farmscape* venue. The Independence schools have award-winning instrumental and vocal music programs and a growing speech and drama program, despite the lack of an auditorium. Locals have come to refer to the high school gymnasium as the gymnitorium. The gifted and persevering music directors have made the most of a difficult situation by utilizing portable speaker and platform systems. As a side note: Almost a year after our *Farmscape* performance, a bond issue was passed to build a new junior/senior high school, complete with a 700-seat auditorium.

We finally picked a museum for our venue. Independence is also the home of the Heartland Acres Agribition Center, an agricultural museum in the Silos and Smokestacks National Heritage Area. The area, which comprises central to northeast Iowa, is one of 49 federally designated heritage areas in the nation and is an Affiliated Area of the National Park Service. On the lower level of the barn-shaped museum, an Amish buggy stands next to a mechanical cow which children love to milk. A small movie theater traces the history of the area in film. There are chicken incubators for the spring babies, milk cans, and bales of straw, not to mention a faint aroma of farm animals from the livestock area next door.

On the upper level are displays of pioneer and modern implements against an encompassing panorama backdrop painted by local artist John Shaffer. Outside a Texas longhorn bull, pigs, goats, cattle, chicken, geese, and ducks live in the summer. "Big Bud," the world's largest farm tractor, is housed in a machine shed along with historically significant tractors, combines, and other implements. A one-room schoolhouse is also part of the exhibition.

Connected to the museum on the east end is an area filled with privately owned antique cars and a rotating display of artifacts from the Buchanan County Historical Society. The far east end of the complex is an event center used for banquets and exhibitions. When we first approached Heartland Acres, the Executive Director generously waived the facility fee and blocked out a section of the event center for the performance.

Cultivating interest

One of our early concerns, funding, was overcome when Ms. Swander, with the backing of the Leopold Center for Sustainable Agricultural at Iowa State University, helped pay for her three-day residency by transferring money from a venue which had been forced to cancel because of a flood. The other half came from a very generous businessperson who wished to remain anonymous.

In a small town, engaged citizens tend to be very involved in a number of activities, events, and sports. LACES, with the leadership of our president Mary Kay Johnson, a legal secretary and active volunteer, persuaded people to add one more event to their calendars. She was the hub for coordinating communications and organizing the various tasks. My son, John, as a staff member of the *Independence Bulletin Journal* newspaper, helped contact and rally participants also. He promoted publicity in the local papers, radio, cable TV, posters, and a web site. Literary clubs, Ladies Musical Society, Red Hats, PEO, Friends of the Library and the Chamber of Commerce spread the word and gave support as well.

We chose the cast from acquaintances and friends who had been active in community activities such as churches, 4-H, Four Oaks, Independence City Council, and civic organizations. The cast members included farmers, teachers, a John Deere engineer, a full-time auctioneer, a priest, a secretary, and a great keyboard pianist (and trained auctioneer as well).

Plowing ahead

We thought we had everything lined up, when we hit a bump in the road. The position of Heartland Acres Executive Director had a change of personnel and our *Farmscape* commitments were not in the files. We were dismayed to learn the event center's auditorium was re-booked with a paying client in the transition. We had to scramble and compromise a little, but together we worked out a new plan. We were given the opportunity for a more intimate setting in the museum.

We were tucked into a corner of the museum's first floor, where we could close the barn doors to the small animal area and hang ordinary bed sheets for a back drop/slide show screen. The ceiling above was the exposed wooden planks of the second floor exhibition area. We were able to borrow risers from the school, corn shocks from the museum and used Ms. Swander's portable benches to complete our minimalist staging.

The pastor of St. James Episcopal church, Rev. Sue Ann Raymond, opened her Fellowship Hall for two evening practices where the cast bonded in conversations and refreshments. During the day, Ms. Swander taught two workshops at our high school and participated in a book signing at a downtown shop.

Reaping the benefits

Then came the grand event. We had a wonderful performance and a great crowd. We had decided to use the PG-13 rated script for our community. It still packed a punch though when Father McManus, a local priest, speaking as Martin, called riding a favorite tractor as good as sex.

Since there is a strong emphasis on conservation in this valley of the Wapsipinicon River, many observations and post-performance conversations centered around those aspects. *Farmscape* was a very effective way to enlighten and encourage people to be better stewards of our soil and environment. It has helped us to appreciate and preserve our heritage and encouraged us to be innovative in our plans for a promising future.

Beyond the immediate event, *Farmscape* sparked further discussion around dinner tables, at coffee klatches, committee meetings, political groups, and clubs. The character representing the packinghouse employee brought up the tragedy of the raid at Postville, Iowa, and the need for government regulations as well as more humane immigration reform. The winery couple reflected a growing interest in new enterprises. And the auctioneer reminded us of the changing structure of family farms to factory farms with new environmental problems.

We became more aware of the Ethanol plant at Fairbank and the growing number of wind turbines. The school board is considering adding a course in agriculture to the curriculum when the new school is built. There was speculation about the on-going establishment of Monsanto and genetic engineering west of town. Crop dusting and expansion of the local airport were topics of interest. The county engineer emphasized the need to repair farm to market roads and bridges. Concern was expressed about the changes in the Buchanan County Amish farmers as they seek jobs in construction to maintain a livelihood and preserve their lifestyle. We became more aware of the traffic pattern of trucks passing through our downtown and the effect on our economy from the many local families in the trucking business, carrying livestock and dairy products all over the nation.

One group visited the corncob processing plant, which collects residue from hybrid seed companies and recycles it for use in green ways. Conservation leaders have continued with programs to plant switch grass and trees and improve drainage systems. Our Master Gardeners enrich the beauty of Independence and model a healthy lifestyle.

Not that *Farmscape* has brought about a revolution, but it has helped fuel positive change and dialogue in our community.

Creating a Celebrity Cast for Farmscape

Vicki Simpson

I was delighted when Mary Swander asked me to write about *Farmscape*. I was the producer for three productions of her docu-drama at the James & Meryl Hearst Center for the Arts. *Farmscape* was my first professional experience with the words, "Sold Out." While I regretted having to turn people away at the door for our first production of *Farmscape* on March, 7, 2009, I was intrigued by its popularity. My role as producer was really very simple—find an impressive cast and let *Farmscape* work its magic under the direction of our state's Poet Laureate, Mary Swander. In fact, my role as producer was so effortless that we even took the cast on the road to perform for the Agricultural Communicators in Education's annual conference in Des Moines on June 7, 2009.

Now, I am not into numerology, but I find it very interesting that all three of our productions took place on the 7th day of a month. This was not intentional, but I love connections like that. I am drawn to odd similarities. I am also intrigued by opposing forces and the energy that they bring to art—be it in visual, literary, or performing arts. *Farmscape* has many such conflicting forces. It is loaded with tensions between early and late twentieth century farming—the beauty of the land, and those who work it—juxtaposed with underlying threats to that beautiful environment and its robust caretakers—vivid contrasts make for the highly poignant, ever provocative underpinnings inherent in *Farmscape*.

As the development coordinator at the James & Meryl Hearst Center for the Arts in Cedar Falls, Iowa, I am responsible for marketing, public relations, grant writing, fund-raising, and public events. I regularly seek out cultural opportunities to further our mission: "To open doors to the humanities and creative arts." When I became aware of *Farmscape*, I knew it would "open doors" and create connections between two seemingly diverse groups—urban and rural—by helping both groups engage in a conversation of importance to all of us. I was attempting to underscore the mission begun long ago by farmer-poet, James Hearst, our benefactor. Like Mr. Hearst, I believe we can all play an integral role in the process of enlightenment and edification essential to human understanding. Particularly in the area of food production—a necessary, but often (at least outside of the agricultural community) little understood element of our lives.

With my marketing background, I also wanted to make sure that our *Farmscape* productions would be well-attended. I was fully aware of the plethora of cultural and entertainment options available to community members. I wondered how I could make our event stand out. I needed a hook. I thought about this and realized that most major entertainment venues offer *stars*. I decided that a *celebrity* readers' theatre was the direction I wanted to go.

Outside of the obvious red carpet magnets that one routinely sees on the national level, what constitutes a local celebrity? That's a rhetorical question, obviously—but play along with me, I think you'll enjoy where this is going. "Celebrity" is subjective; it means different things to different people. For me, as an MA student at the University of Northern Iowa, professors were my celebrities; they embodied the intellectual stamina that we students aspired to—and they did it (most of them) with a certain amount of panache.

Next, I looked at the farming community and how it has melded with higher education—which is how I got the idea to invite Agricultural Extension members to be a part of our production. Then I began to contemplate the significance of media to both rural and urban residents. I thought of newspaper, television, and radio personalities—the men and women who disseminate information to both sectors on a daily basis—typically with a great deal of charisma. More stars!

Additionally, I figured that I needed to include some well-established and beloved veteran community theatre actors and place them on stage. As well, I considered the importance of energy and the environment to both the agricultural and municipal camps—I wanted to augment those sensibilities with a representative from city government—there are stars abounding all over city government! Lastly, I topped the cast off with a dairy owner/operator who is not only the *face* of his family's wholesale/retail dairy chain, but he'd also been acting since high school. With this I knew I had the eclectic mix of personalities needed to attract a diverse audience.

In a nutshell (or two), that is how the idea for the Hearst Center's celebrity reader's production of *Farmscape* was conceived; a model that, according to Mary Swander, has been replicated with great success in subsequent productions of *Farmscape* throughout the region. Aesop indeed had it right: the first in the market owns the lion's share—I think this celebrity model worked well for our productions of *Farm-*

scape. I highly recommend it for your community. A simple phone call to potential cast members was all it took. Each of the following members enthusiastically embraced the *Farmscape* idea from the onset, defying the typically held belief that celebrities are "unapproachable."

Our cast of gracious celebrities included: Tara Thomas, KWWL Iowa's News Channel 7 Anchor/Reporter; Melody Parker, *Courier* Communications Editor; Jim Coloff, Mix 93.5/1650 The Fan, Owner/General Manager; Brent Hansen, Hansen's Farm Fresh Dairy Owner/Operator; Kamyar Enshayan, Director, Center for Energy & Environmental Education, University of Northern Iowa, City Council Member, City of Cedar Falls; Dr. Jim O'Loughlin, Associate Professor, University of Northern Iowa Languages & Literatures; Dr. Grant Tracey, Professor, University of Northern Iowa Languages & Literatures; Dr. Kyle Troyer, Upper Cervical Chiropractor; Al Ricks, Iowa State University Regional Extension Education Director; and Jaime Reyes, Owner/Operator of MonteBello B&B Inn, Iowa's Prairie Hacienda.

These high-profile readers imbued their characters with warmth and originality, effectively giving the audience a rare glimpse into the real-life struggles of Iowa's farm families. Tara Thomas spoke from the heart in a recent interview:

> "The *Farmscape* experience was twofold for me. Not only did it get me out of my comfort zone for my first time by doing readers' theater, it also connected me to my farming roots as both sets of grandparents lived and worked on farms. As a city girl, I now appreciate the challenges so many Iowans faced during the farm crisis. *Farmscape* enhanced my role as a news anchor and reporter in that

I can better articulate agriculture stories with a grasp of where we've come from."

Tara's news broadcasts have been staples in most farmers' homes since 2002 via the nightly news on KWWL. Tara played the role of Kristi. This was a huge draw for our audience. It's not every day that the agricultural and urban communities have the opportunity to get close to a news anchor that virtually comes into their living rooms, kitchens, and even their barns with the latest information—and rarely do they get a glimpse of a celebrity performing such a unique story!

The same can be said for radio station owner Jim Coloff. Jim played Kristi's husband, Randy, as well as Max. Jim was raised on a farm and so the *Farmscape* experience was uniquely personal. As a youngster growing up on a farmstead he learned the value of hard work, and what goes on behind the scenes in the growing of the food we put on our tables. Jim said it was a tremendous experience to be able to, "get back to his roots," in *Farmscape*.

As a child Jim Coloff was also very active in 4-H, this was also close to cast member Al Ricks' heart. As Director of the Iowa State University Black Hawk County Agricultural Extension, a position he held for many years, before retiring to spend more time with family, Al had been actively involved in farming and 4-H as well. This reverence for agriculture came out beautifully in his portrayal of both Donald and Jaime.

When Al switched gears from playing Jaime in our first production, to Donald for our second production, the real-life Jaime Reyes, Owner/Operator of MonteBello B&B Inn, was gracious enough to step in to portray himself. It was delightful to have Jaime join the cast in the role he inspired. In an interview Jaime explained that being in *Farmscape* really opened his eyes, "I learned that agriculture is not only to plant a seed and watch it grow. It is far more than that. It is a very complex social fabric that involves and affects many lives." When asked what he thought of using celebrity readers, Jaime replied, "casting stars does provides some glamour to the show. However, it is more important to have excellent readers who can truly identify the characters and make them their own."

Finding excellent readers who could identify with their characters, while providing a certain iconic allure was my two-fold mission. I was fortunate that I found local celebrities who were willing to give their time and energy. Another was Dr. Grant Tracey who deftly played the lead of Martin, a farmer faced with losing everything he has worked for his whole life—at his own auction.

Celebrity reader, Dr. Jim O'Loughlin, played multiple roles in all of our productions of *Farmscape* too; playing Mike, Nate, and Max's son-in-law. Jim was able to morph into a new character by a simple turn of the hat, which was quite amazing to watch! When asked about his experiences in *Farmscape* Jim explained:

> "Performing in *Farmscape* was a unique experience, and only Mary Swander could have worked the magic necessary to make it happen. Because people knew that *Farmscape* was a work of non-fiction, audiences recognized their neighbors and friends, as well as their own experiences, in the drama, and after the performances they often wanted to share their own stories with us. And, because sometimes we were performing right alongside the actual people who were originally interviewed for the project, the distinctions between subjects, actors, and audiences

started to blur together in what started to feel like one big community."

Liane Nichols, known for her community theater roles throughout the years, found her part as Daphne (Jaime's wife) to be quite singular on a number of levels—one being the fact that the real Daphne was in the audience at our productions. Liane channeled Daphne with sensitivity and aplomb. Having Daphne Reyes observe her performance imbued her character with a delightful sense of multiplicity that further enhanced the multi-voiced components of the whole *Farmscape* experience.

Like Jim Coloff, Liane also felt there was, "a warmth among the cast and between us and the audience that was gratifying." Due to a situation beyond her control, Liane was unable to play the role of Daphne in subsequent Hearst Center productions, which gave me the opportunity to step in and lend a hand. I was deeply impacted by the presence of some of the real-life people that inspired the roles in *Farmscape* as well. A contingency of graduate students, who helped Mary write the play, were in the audience at all three productions. Talk about doubling, doppelgangers, and blurring of boundaries!

Melody Parker's character, Lonna, was also in the audience, which made for a powerful reading of her role as well. As a master gardener and longtime Lifestyles Editor for the *Waterloo-Cedar Falls Courier* Melody provided a high level of understanding and credibility to her portrayal of Lonna.

Cedar Falls city council member, Kamyar Enshayan, discussed his take on playing Donald in the Hearst Center's premier production, "I enjoyed it very much. We drive by it all [the farms and the land], but by participating in *Farmscape* we directly experience what all is going on in our own state."

Driving by, but being unable to wrap our minds around certain elements of the food chain is something all too intimately understood by dairy owner and operator, Brent Hansen, who played the foul-mouthed, racist, meat processing plant worker, Jon. Brent said he loved playing the role of Jon because of the, "shock factor—everybody is so complacent. The city we live in is so protected—that's not how it really is." He went on to describe how his, "racist, rough-around-the-edges character is the provider for all of the people." This destabilization factor—which becomes a revolution, of sorts—is taking place from the margins (where all great revolutions begin) and it is part and parcel of the magic intrinsic to *Farmscape*.

Farmscape takes everything you have ever thought about the food production industry and turns it on its head making for an unforgettable conversation.

What Is Lost

J. Harley McIlrath

My uncle Stan stands on stage—he's Mike, the successful farmer. He stands on stage, and he's Mike, but he looks like Uncle Stan because Mary Swander had us put together our own costumes, and Stan's worn what he would have worn anyway. What's the difference? He's already a farmer. "I farm 1,700 acres," Stan says–Mike says, "My father farmed and my grandpa did too," and whether he's speaking for Mike or for himself, it's true. He farms more than 1,700 acres, and the McIlraths have always farmed. Like Mike, Stan used his dad's machinery when he started, and he farmed his dad's land. But then, Stan says–Mike says, when he started farming, he had nothing to lose.

That's wrong.

Later, he says it again. "I really had nothing to lose because I really didn't have anything going into it. I didn't have any down payments on anything purchased…[You] take zero away from zero and you still have zero."

Mike's talking about his bank account; he's talking about the business he went into—and that's where it becomes clear that my uncle Stan is only mouthing the words of a character, because Stan had everything to lose when he took over his father's farm. His was not a business decision; it was the fulfillment of a way of life. There was no decision. Being a farmer was who he was, who he had always been. Stan was a farmer before he was born.

You take the farm away from the farmer, you take away his soul.

I'm on stage as Randy, the guy with the hog confinement. "I'd say it's roughly 100 foot by 300 foot," I say, talking about the lagoon that holds the liquid hog shit. The script says "feet," but I say "foot" because that's how it rolls off my tongue. That's how my dad would say it, foot instead of feet. That's how I would say it if I hadn't gone off to college for twenty years and wound up selling books for a living instead of farming. "It's 15 foot deep," I say, and it's natural. I hear my voice saying it like I'm listening to the other me talking, the me I used to be, the me I'd be if I hadn't left.

Why am I up here?

Mary's had us put together our own costumes, and I didn't try much harder than Stan did.

"What're you going to wear?" my wife says. "What's a hog farmer look like?"

She's a city girl.

I go to my closet and I take out the plaid shirt I never wear, button down front, no buttons on the collar. I never wear it because it's the kind of shirt the boys'd wear to the Co-op meeting, to the West Side Diner for coffee, to the sale barn if they cleaned up. It's a hand-me-down from my dad. I grab a pair of jeans, brown leather boots—they should be Wolverines, but I've got Doc Martens. I lay it all out on the bed. Then I get a pair of scissors and I cut the sleeves out of the plaid shirt. I walk over to my parents' house next door and ask to borrow a cap.

"Which one?" my dad says.

"I'm Randy, the hog farmer," I say.

"What's that mean?" my dad says.

"Give me one you don't want back," I say.

Of course, he gives me a nice one…because I'm going to be on stage.

I go home and I put it all on, my costume. I stand in front of the mirror.

"You look like Eddie Vedder," my wife says.

Grunge. The godchild of Neil Young—seed corn caps, flannel shirts, and electric guitars.

"I look like the kid who showed the Reserve Champion Market Pig at the 1974 Poweshiek County Fair," I say.

My wife doesn't know what that means. It means I look like myself. I look like everything I chose not to be.

Stan and I are only four years apart. He's my uncle, but he's only four years older. When we were boys, my house was right next to his, the yards separated by a row of peonies. In summer, Stan and I rode our bikes out the dirt lane that ran from the barnyard west along a row of honeysuckle and then turned north along the web fence until it reached the pasture gate. When we were older, we drove tractors up and down the lane, pulling wagonloads of corn, beans, and racks of hay. Up and down the lane we went until the packed tracks the tires made turned to powdered dust. Out the lane we went to mow hay, to check cows. When it rained and it was too wet to be in the field, my grandfather threw a bucket of staples into the bucket of the little Ford tractor; he threw in a hammer, fencing tool, cinch, and sledge. He threw a roll of barbed wire in the tractor bucket and a couple of spades. Off we'd go down the muddy lane, Uncle Stan, Grandpa and Grandma, my aunts, and me, wearing five buckle overshoes. We'd follow the fence line with the little Ford, fixing fence—tightening the wire, replacing staples—and my aunts would roam with the spades, digging thistles from the fence line, from the pasture, from the sloughs.

When we planted at my great-grandmother's farm, Stan and I played in the field, or in the fencerows. We might play in the yard around the house, but we did not go in the house, and we didn't stray far from the field in case we were needed. We were boys. Most often we sat on the fenders of the tractor—my grandfather turned in the seat, one eye on the field ahead and the other on the planter behind—or we followed the planter with my father, digging in the soil with his finger to check the depth of the kernel, the distance between the seeds.

In the house, my great-grandmother sat at the kitchen table and watched the men through the window. My great-aunts, Ruby and Mabel, peeled potatoes and set the table for dinner. My mother helped them, and my grandmother hovered between the work in the kitchen and the work in the field, ready if a shout came from the field that help was needed. My aunts would be there. Before long, a car would pull in the driveway, and my great-aunt Lizzie would get out, and Uncle Alec, drawn home by the work, because this is how it had been since Lizzie was a girl. This was her home, and the men were planting. An old woman, she would scurry from the car into the house, take up a paring knife and a potato, her excuse to sit at the table, to be a part of the family.

An old man, Alec would saunter through the gate and out to the field to be with the men, even if, for him, being with the men meant leaning against a fence post at the edge of the field—and perhaps he would step away from the fence post to catch a seed bag taken by the wind, roll it, and tuck it under his arm. He would have done his bit. He would have a place at the table.

When dinner was ready, if we were nearby, Stan and I would be sent to the field to tell the men, and when the men had walked to the house, bringing Uncle Alec with them, they would roll their sleeves above their elbows and wait their turn at the wash basin. Mabel and Ruby would bring the food to the table—roast beef on a platter, boiled potatoes, beef gravy, green beans, bread and butter, canned peaches—and when my grandfather and my father had scrubbed their hands clean, scrubbed their forearms to the elbows, Alec, too, drying his hands with a towel, they would come to the kitchen and find their places at the table. We crowded around the table then, Stan and I, my aunts, my father and mother, my grandparents, the great-aunts and great-uncle, Great-Grandma.... The family ate.

This is what I left.

But it's true, too, that if you don't leave a place, in time the place leaves you. My grandparents' house, the house in which Stanley lived as a boy, is gone now, and gone is the little house in which I lived with my mother and father. Gone are the peonies and the yards. The lane down which we rode our bikes and drove tractors is gone, and gone the honeysuckle, the web fence and the pasture gate. Gone is the pasture. Gone is my great-grandmother's house, gone the yard and the fencepost on which Uncle Alec leaned. Uncle Alec himself is gone. Great-Grandma, the great-aunts, Mabel and Ruby and Lizzie, my grandparents—all are gone.

I miss crowding in with my family—I miss the closeness, the be-
longing. That is what is lost, and that is why I'm up here pretending
to be a person I might have been, pretending to be a person I chose
not to be. I miss my family. I miss my uncle. I miss waking up in the
morning knowing that our day will be spent together. I miss being
with Stanley, pressed tight between my father and grandfather in the
cab of a pickup, my grandfather at the wheel, my father with his arm
out the window. We're headed to town for a part, or we're headed to
the sale barn—we're headed somewhere—and the wind is blowing
through the cab and the dust is roiling out behind us.

So here I stand on this stage, because here is my uncle. We're together, doing this together.

"I'm Mike," my uncle Stan says. "Number 51."

"I'm Randy," I say. "Number 20."

The Backstory of the "Chicken-Plucking" Latino/as in Farmscape

Claudia Prado-Meza

Latino/as figure into *Farmscape* through two very different characters. Jaime, the Bed and Breakfast owner, is a Mexican immigrant, a middle class highly educated entrepreneur with an engineering degree. He designed and built his beautiful home, had the resources to fight the city over water rights, and the capital to set himself up in business. He wants to establish a Latino cultural institute to preserve his own heritage and educate others about the visual, musical, and literary arts that he loves. Jaime is adamant that his community should be given the opportunity to enjoy the finer aspects of Hispanic culture. At one point in the play, he says:

> We plan to build a Latino cultural institute on our land.
> We're going to bring arts, artists, from overseas—exposi-
> tions of painting, music from Latin American countries.
> To create a better understanding between both cultures.
> Because there are other Latinos besides the chicken-pluck-
> ing Latinos! Latinos that are cultured, you know?

On the other hand, Jon, the play's slaughter house worker, knows little culture and is surrounded by "chicken plucking" Latinos. Jon speaks of these Mexicans with unease, lumping the Bosnian immigrant workers into the same prejudicial stew:

> My co-workers? I had suspicions with the Bosnians and the Mexicans, you know, feelin' unsafe 'n shit. But they're just makin' a livin'. I guess if I'm bein' honest, um, yeah, uh, I'd rather be by more, more, uh, workers like me. Who wouldn't?

So, how did Jon, a young, white Iowan get thrown together with Bosnians and Mexicans at the meatpacking plant? Why and how did Latinos come to Iowa and other U.S. states? What are the socio-economic factors that play into the changing demographics of small Midwestern cities? What's the backstory behind the diversity of laborers at slaughterhouses like the one Jon works at?

Jon does not name his meatpacking plant. It could be one of a dozen such plants in the state. Each plant is different, but the stories of the workers are universal.

As a sociologist, I have studied Latino/a immigrants in Marshalltown, Iowa, where they came to work at the Swift plant. The tales of the following two immigrants—Magdalena and Manuel—are representative of Latino/a working class laborers in Iowa. They symbolize the challenges and circumstances of "chicken-plucking" Latino/as–Mexicans that Jon may very well have known and had as co-workers.

Magdalena was born in the United States to Mexican parents who came from Villachuato, Michoacán, México, to work at the JBS Swift meatpacking plant located in Marshalltown. The facility slaughters around 19,000 hogs a day. Latinos started arriving to this Midwestern city at the end of the 1980s and beginning of the 1990s and were then about 1% of the city population. Now, almost 25% of the total inhabitants of Marshalltown are Latino, drastically changing the demographic landscape of the city. Latinos in Marshalltown are not only working at the meatpacking plant, but they are also activists, students, and business owners.

Manuel has lived in Marshalltown since June 1989. He and his family are also from Villachuato. He and three of his friends were the first Villachuatans to arrive in the city. Manuel was granted amnesty through the Immigration Reform and Control Act of 1986, commonly known as IRCA. He was a migrant farm worker who used to travel around the United States following the field seasons. Most of his time was divided between the states of Washington and California. If Manuel couldn't find fieldwork in Washington, he turned to factory jobs. At the beginning of 1989, a friend invited Manuel to come to the Midwest to detassel corn during the summer. Manuel decided to give that a try. He traveled to Lexington, Nebraska, where a large community of Villachuatuans was already living. There, he heard about work in the Marshalltown meatpacking plant that would allow him to get off the road and settle down.

At the beginning, the Marshalltown idea did not sound very appealing to Manuel and his friends, but then they learned that it would be rather easy for them to get a job at the plant. The job had several advantages. It was year-round work. The men did not have to speak English fluently, and they could afford to own a house—for many of them, for the very first time. However, Manuel and his friends had to be willing to endure the long shift hours, the cool temperatures at the plant, and get used to the Iowa weather. A few months later, after accepting work at the plant, the *majordomo*, or manager, asked Manuel to invite friends and relatives to come to Marshalltown to work. The plant was about to open a new shift. And in just 20 years, a transnational community of Villachuatans was living in Marshalltown, Iowa.

It might seem that Villachuatans arrived in Marshalltown by chance—a dart thrown at a map. In reality Villachuato and Marshalltown have many similarities. Marshalltown is surrounded by corn and soybean fields while Villachuato is encircled with vast fields of corn, sorghum, wheat, and fava beans and with billboards advertising Pioneer, Dekalb, and Syngenta. The land in Villachuato is productive, and Villachuatans are proud of the quality of the produce that they grow, but inputs are expensive, and small-scale farm harvest profits very low. With little income, many Villachuatans do not have enough to eat, and feel forced to immigrate to the United States. "*El norte nos quitó el hambre,*" one of my Marshalltown interviewees said. "*The U.S. took away our hunger.*"

Changes to the Mexican constitution and federal food and agricultural policies have not helped Villachuatans. On January 1, 1994, Mexico signed the North American Free Trade Agreement (NAFTA) opening the way for *ejido*, or communal land used for agriculture, to be sold to landowners. Mexican agricultural production and subsidies increasingly centered on large-scale farms, factory-type livestock lots, and capital-intensive food processing. Little support was given to small family farms like those near Villachuato.

Constant migration has also reshaped the Villachuatan's landscape. Young men started to leave wives, daughters, and mothers behind. They sent money back home that their relatives used to build two-story concrete houses. Even the *ejido* benefited from the remittances.

The commissioner of the *ejido,* or *comisario ejidal,* stated that the Mexican farmers used the money to buy fertilizers, herbicides, and equipment for plot leveling on their specific parcels of land.

The *comisario ejidal* explained:

"Si, los hijos mandan dinero para acá, y los papás invierten mas en sus casas y en las parcelas. Si, [el ejido] ha mejorado. En aquel entonces [antes de que empezara la migración] estaba muy mal, y actualmente tenemos unos sorgos y unas tierras bien bonitas."

"Yeah, the children send remittances and the parents invest that money in their houses and in their land. So, yeah, the ejido has improved. Before [Villachuatans started migrating], it used to be really bad, and nowadays we have very pretty lands and sorghum."

Agriculture is vital to both Marshalltown and Villachuato. Food production shapes traditions and family and community values in both cities. The local job market doesn't have much to offer youth in either town, so young people leave to find better opportunities.

In *Farmscape,* Jon voices the frustration and depression of the local Iowa youth who are left behind to work in the meatpacking plants:

My body hurts too bad to play sports—I sucked at football, you know, but I liked it—and uh… m…No, no injuries or shit. Just tired. At training, yeah, that's right, they tell you to take a shitload of Tylenol or whatever days before you start. You're doing the same damn thing everyday, which is crap for your body. I took hot baths every night for about a month 'till I got used to it. You know, bring a TV in the

bathroom. It's not so bad. Everybody I know is graduatin'
and getting' jobs and movin'… I'm… And I'm… Yeah. I
can't get away from the packing plant.

Yet both communities are resilient. Villachuatans do not forget
about their hometown. They go back to visit when they have the opportunity. But they also get involved with issues in Marshalltown. For
example, Magdalena plans on going to law school to improve the lives
of the younger generations of Latino/as in her community.

Marshalltown is also changing to adapt to its new reality. Schools
are now embarking on bilingual programs. Local leaders and food
lovers are inviting immigrants and refugees to be part in a developing
local food system, a system that might offer Latino/as and locals alike
a better job with a better wage, so that characters like Jon might not
feel so "stuck."

I'm fine with the hard work. It's just… how much do I
make a year? Don't ask me that. I don't make enough to
take a goddamn vacation. It's hard work, you know, uh,
has-to-be-done type shit, but I can't ever afford to get away
from it.

The locally developing Marshalltown food system offers Villachuatans the opportunity to create local jobs, to give young beginning
farmers a chance to grow food organically and sustainably. Thus,
citizens of Marshalltown—both local and immigrant—are starting to
understand and exercise their power to create an ideal community,
and good places for kids to grow up.

Where Are Women's Voices in Agriculture?

Leigh Adcock

Worldwide, women produce more than 80 percent of the food we humans eat, in spite of owning less than two percent of the land it's grown on. Women everywhere take responsibility for feeding their families and communities, and struggle to build local and regional food systems that make healthy food available to everyone. In spite of operating only 14 percent of U.S. farms, women are the primary drivers behind the booming healthy food and farming movement—as producers (up 30 percent between 2002 and 2007), consumers (responsible for a majority of household purchasing decisions), and most staff at non-profits that support the movement.

Nurturers by nature, women monitor and strive to improve the health of their families and communities, and have been among the first to speak out against the environmental degradation and harmful health effects of the U.S. system of industrialized agriculture and food production. As producers, women are more likely to operate small-scale diversified agricultural farms than conventional grain farms. The 2007 USDA-National Agricultural Statistics Service Census of Agriculture data shows that women's farms tend to be smaller, but women are more likely than men to own all of the land they farm. At the Women, Food, and Agriculture Network (WFAN), we hear from women all over the world who began raising food to feed their families and communities because they refused to participate in the corporate industrial system that currently dominates U.S. food and farming.

"My primary interest is getting my family to eat a diet that's natural to the human body, away from processed foods," wrote one woman in a 2011 WFAN member survey. "We are also trying to be better stewards of the earth for our children's future and posterity. My interests are also being passed on to our two daughters, one of whom already wants to be a farmer when she's an adult."

In the Midwest, where *Farmscape* was born, there is a striking contrast between the type of farming most women prefer and the dominant agriculture on the landscape. Iowa is a Corn Belt state. Most years, Iowa is the number one producer of hogs in the country—yet the state is a net importer of food. In response to this dichotomy, a growing number of Iowans are creating community-supported agriculture farms (CSA), farmers' markets, urban community gardens, and edible schoolyards. The local food movement is growing rapidly, even in the number-one commodity agricultural state in the U.S. With this growth has come a great deal of conflict—usually of the polite, Midwestern variety, but conflict nonetheless.

Lonna and Kristi, two women characters in *Farmscape*, perfectly illustrate the tension between feed vs. food farming. Lonna grows food to feed families in her community. She and her husband Joe make a living on an acre-and-a-half of land near an urban market. She speaks movingly about caring for her vegetables, about providing them with what they need to thrive (some plants like "cool things around their roots.") She refuses to eat confinement pork and points out that all meat-eaters should participate in at least one slaughter in order to appreciate the moral dimension of taking a life for food. Lonna also admits that "money isn't one of the biggest things we get here." In addition to subverting the agricultural paradigm with her small-scale food farm, Lonna opts out of the first rule of capitalism: the more cash you earn, the happier you must be.

Kristi is the most poignant character in the play. Her story is repeated in thousands of households across the Midwest. Farm wives give up a great deal of agency in order to fit into the prevailing agricultural model. As a young wife and mother, Kristi was told by her in-laws that she would be moving from her home to theirs, in spite of the enormous amount of love and work she had poured into hers. Years later, she still resents the move. Kristi clearly feels the stress of managing an industrial hog confinement, both moral and practical. She points out that her husband was the subject of a university training video on manure confinement management, but Kristi admits that the concentrated odor in the building can kill both animals and people. "That's livestock," she says.

But does she really believe that cavalier statement? As a wife and mother, Kristi is responsible for maintaining the physical and emotional health of her family. Even though she may feel the confinement model is unnatural for both people and animals, she feels bound—as many Midwest farm wives do—to defend the choices of her father and husband. When the farm crisis of the 1980s came, she says, "... you had to get bigger or get out and move to town." Tellingly, Kristi also points out that contract livestock was her family's way to hang on to their farming life. "Some of us [do] whatever we can to hold on," she says. "It's just not in [our] blood to work in town." She and her husband Randy are keeping the farm they own so their son can farm it, "after this contract is up."

One of the most heart-breaking consequences of the schism in Midwestern agriculture is the divide that has sprung up between women invested in the conventional system and those forging careers in local and regional food systems. Women are passionate on both sides of the issue, and arguments can grow heated when values come under fire. An ironic twist is the recent creation of the Common Ground campaign, funded by corn and soybean groups, to "put a softer face" on conventional agriculture and food production by sending farm wives into supermarkets to reassure consumers that their processed food is safe. Their website purports to "debunk myths" about issues of food safety, environmental quality, and livestock production. A team of women bloggers and spokespersons passionately and vocally defend the dominant system, which has come under fire from every side over the past decade.

The reunifier will be found at the intersection of economy and health. According to the 2007 USDA Census of Agriculture, the number of women farm operators grew 30% between 2002 and 2007.[1] Women all over the country are creating farm and rural businesses to provide or augment the family income while also providing for the health of their families and communities. At the same time, frightening statistics about the rise in childhood obesity and other diet-related diseases are pushing more and more mothers into the aisles of the whole food stores, and away from industrial "Franken-food." The combination of this market pressure and the natural propensity women have to tend living things (children, gardens, and livestock) is changing the face of food and farming in the U.S. For instance, sales of organic and natural products are claiming a growing share of the market; home gardening is booming, and—most encouragingly—public pressure is on the rise to create food and farm policy that rewards environmental stewardship rather than bushels per acre.

But women's voices in agriculture are still muted. More than any other major industry, agriculture is policy-driven. The products we raise, the markets we sell them into, and the methods we use to grow them are dictated by federal farm policy. In turn, farm policy is the product of an uneasy tension among legislators, the commodity groups and agri-business conglomerates that lobby them, and citizen groups advocating for public and environmental health, and rural community development. An Environmental Working Group report issued in early 2011 revealed that only 3 of 225 board members of the five largest U.S. commodity organizations were women. Women represented only 22.1% of all state senators, 25% of all state representatives, and less than 17% of the U.S. Congress in 2011.

"That's livestock," Kristi says. That's farming. That's progress. But deep down, I believe she knows better. We all do.

WFAN was born in 1997 when a handful of women farmers and food activists looked around and realized women were not at the decision-making table at any level. In 15 years, our membership has grown along with the healthy food and farming movement to include farmers, consumers, advocates, students, nutritionists, health care professionals, and policy-makers. We are urban and rural, range in age from 16 to 85, and come from a wide variety of professions. Our common goal is to amplify women's voices in the ongoing agricultural debate. Our mission is to "link and empower women to build food systems and communities that are healthy, just, sustainable, and that promote environmental integrity."

"Women approach farming differently than men do," wrote another survey respondent. It's a simple statement, with the potential to transform everything around us. How different could farming and food look if we made up 51% of the decision-makers, as we represent 51% of the population? How much healthier could our food, families, and communities be? It's time for us to make our voices heard, from the farmhouse to the White House.

Note:

[1] United States Department of Agriculture, National Agricultural Statistics Service, *2007 Census of Agriculture*, 2009. http://www.agcensus.usda.gov/Publications/2007/index.php

Swearing in Public

Jason Arbogast

My English: 557 course created *Farmscape* in 2007 a little over four years ago. I'm still involved in the project, officially making the class the longest I've ever taken.

Our dress rehearsal was on a cold February night at a small coffee shop in Ames called Café Diem. I'd done many open mic nights in coffee shops, so I was used to speaking over the espresso machine that kept trying to drown us out. We were surrounded by audience members—both in front and behind us. This set-up was a little different, but not a big deal. The question was, why was I there? I'd already gotten my grade for the class which had ended back in December. So why did I perform in this show?

Simple. I wanted to swear in public.

I turned the bill of my seed cap forward on my head during *Farmscape* performances when I read the part of Nate, a Monsanto higher-up whose family used to farm. I turned the bill of my seed cap backwards when I read the part of Jon, a guy who had graduated from high school and gone right to work at a processing plant ripping out pig guts, which he had no problem describing, graphically. These two men couldn't have been more different. Nate is college educated and chooses his words carefully, fully aware that what he says will, in all likelihood, wind up out in the community. He's honest and truly believes that what he's doing is helping the farmers who buy from his company, but he's also political, putting Monsanto in as positive a light as possible. Jon is honest, too, but in a blunt, unvarnished, swearing, first-answer-that-comes-to-mind way. He can come off as racist depending upon who's reading his part. But I always got the feeling he was simply nervous about people of other ethnicities. He hadn't been exposed to them before he began working at the meat packing plant.

I interviewed the "real" Nate for the playwriting class which is why I wanted to read his part. Some *Farmscape* male actors had to take double parts for the production. Jon's role couldn't be confused with Nate's, so they made a good pairing. A Nate/Jon double role also illustrated and contrasted two sides of the corporate world of farming: manager and worker.

And, y'know, I got to swear, loudly and frequently, in public.

I knew how to read Nate. I'd made sure to keep a copy of my interview to use as a reference. It allowed me to get down some of his mannerisms and speech patterns without too much difficulty. Jon, however, was a problem. All I had to go on was what the script contained. It would've been easy to make him into a kind of cartoon character, all bluster, high energy, and profanity, or, worse, some kind of racist caricature, but that wouldn't have helped the play have the impact we wanted and, frankly, it would've been insulting to Jon.

So, what was I going to do? Well, the answer was pretty simple, actually, I thought of my dad.

I grew up in a small farming community, a lot like the ones *Farmscape* depicts, just south of Battle Creek, Michigan. My family wasn't a farming one, though. I knew a few farming families, but that was it. My grandpa and my mom did have small gardens, but that's a long way from farming. No, my family was the other working class type common to my area: blue collar. My grandparents worked in the cereal factories and my dad and uncle worked in an auto parts factory.

I didn't necessarily know farming and farmers, but I did know factory workers. And Jon and my dad have a few things in common. Both have more stories than you can ever listen to. Both are, at their hearts, genuinely nice and friendly people. And both have the sort of mild, learned racism and tendency towards swearing that makes the rest of us squirm a bit whenever they start to tell a joke. I had to channel my dad's edgy speech patterns—but in a good way. Luckily, or unluckily, I'm fine with saying things that make people uncomfortable. And while this includes swearing in public and saying mildly racist things for a part, it also includes the heart of what *Farmscape* is: saying some things that make people uncomfortable.

And, if I get to do it while swearing, all the better.

I'm making a big deal out of the profanity, I admit it. It's something you don't hear a lot in public. It isn't polite, but neither is the message of *Farmscape*. Screw polite. Politeness only gets you so far. After that, you have to be honest, no matter how awkward that might make people feel.

Yes, now is when I pull out my soapbox.

I'll be the first to admit that I wasn't comfortable with a lot of what *Farmscape* had to say. Just like everyone else I grew up believing that more is better, that science knows what's best, and that progress means always moving forward. I still want to believe these things, but they're not true. Problem is, not many people want to say anything about it. Wouldn't be polite.

As Jon's story points out, more isn't always better. Just because we can kill and process animals at an inhuman rate doesn't mean that we should. Not only are safety and accountability lost in the process, but a little bit of our humanity is, too. I'm not saying we should all become vegetarians, though that is a viable option. I am saying that the mechanization and the assembly line-like fashion in which we butcher animals is cruel on a level that most of us don't want to know about. If we did, we might be forced to actually think about where our food comes from, and just what corners have been cut to make it so cheap.

Nate's story, as positive as he tries to make it sound, shows just how arrogant science tends to be when it's trying to make things better, and how corporations don't like to mention the negatives of what they do. For all his pride in how Monsanto is helping farmers and helping the environment, he doesn't say anything about how Europe still doesn't accept genetically modified crops. Europe considers them unproven. Nor does Nate say anything about the level of groundwater contamination caused by the overuse of pesticides. *Farmscape* does.

Farmscape also points out that we should be looking backward for solutions, as well as forward. There's wisdom in the old ways. But this goes contrary to what we, as a youth-obsessed, ever-moving society, tend to think. Monsanto, and therefore Nate, who serves as their proxy in the play, embodies this notion of constantly pushing at the boundaries, never really looking back, never really worrying about the consequences. This is a corporate mentality, not a human mentality. The Great Recession showed us what that kind of mentality in the financial and housing sectors does. I'm terrified of what the farming, and most likely ecological, equivalent to the 2008 financial sector collapse could be.

So, for the past four years, I've been more than happy to continue working with *Farmscape*. Its ability to show people the dangers, problems, and possible consequences of our society's current farming narrative is vital. And *Farmscape* lays it all out in a non-confrontational, sit down, let-me-tell-you a-story sort of way. It's like getting together with a few of your neighbors and talking about what's going on in their lives, rather than being lectured by a bunch of people you don't know or trust.

And, if I get to swear in public a bit, I'm not going to complain.

An Agricultural Alternative

Francis Thicke

On the dairy farm that my wife Susan and I own and operate in southeastern Iowa, we base our farm's design and management on principles of ecology. We offer an alternative to livestock confinement operations depicted in Mary Swander's play *Farmscape*. In the play, Kristi and Randy tell of their hard times during the Farm Crisis of the 1980s. They hung onto their family farm through its conversion to hog confinement. Max speaks of simply accepting the odor of CAFOs, or Concentrated Animal Feeding Operations, near his vineyard. And Jon details the gruesomeness of working in a slaughtering house that processes meat grown on contract.

It doesn't have to be this way. Modern livestock production systems can be designed and managed to mimic the ecological processes that created the diverse prairie and its productive soils. And these systems can be much more energy-efficient than current industrial animal production methods. The key is to find ways to harness the energy, efficiency, and organizing power of nature's ecology. Modern scientific understanding of ecology provides insights into the design and management of these systems.

When bison herds grazed the tall, deep-rooted prairie plants, they re-deposited their manure nutrients back to the soil from whence the plant nutrients had come. And their grazing activities stimulated regeneration and robustness of the ecosystem. After grazing, the shortened prairie plants had excess root mass for their reduced above-ground leaf mass, so the plants sloughed a portion of their roots into the soil. As the plants grew new shoots and leaves above the ground, they also grew new roots below the ground. The root mass that had been released into the soil after the bison had grazed the prairie plants became food to sustain soil microorganisms and produce humus (sequestered soil carbon). Repeated grazing cycles of the roaming bison herds increasingly added to the soil's fertility, productivity, and organic matter.

Susan and I moved our dairy farm several miles away to its current location in 1996. The land we moved the dairy to had been under continuous corn and soybean cropping for many years previously, and the farm had no buildings we could use for the dairy. We had to design and build the dairy from the ground up.

The farmland had been cash rented from an absentee landlord before we bought it, and conservation had been neglected. The land on the farm is rolling and steep in places—with some hills exceeding 15 percent slope. Most of the field's grass waterways were gullied, with the deepest gully nearly four feet. The configuration of the landscape did not lend itself well to contour farming, so row crops had been planted up and down some of the hillsides. In some areas of those hillsides, all of the topsoil was gone. The surface soil—what soil scientists call the "A horizon"—had been lost to erosion.

We repaired the gullies and planted the cropland on the farm to a mixture of grazing forages, consisting of grasses and forbs, including legumes. All the buildings of the original farmstead had disappeared (except an old wooden corn crib). We built new facilities, including a milking parlor, cattle barn, on-farm dairy processing plant, and a house.

We divided about 120 acres of the new pasture into 60 small pastures—called paddocks—using simple, low-cost electric fencing materials. This paddock system allows us the flexibility to manage the location of the grazing cows at all times. We can then optimize the productivity and nutritional value of the pasture forages. We manage the cows to mimic the ecological effects of bison herds roaming the prairie.

We milk about 80 cows twice a day on our farm. During the growing season, we turn the herd out to a new section of pasture after each milking. Normally, we give the milking herd half of a paddock—about one acre—to graze after each milking, twice every day. We can quickly and easily subdivide paddocks into any size increment using portable fencing materials.

In addition to the milking herd, we have two other groups of grazing dairy cattle, each rotating through paddocks in separate areas of the farm. One other group consists of cows that are in the dry phase of their production cycle (each cow calves annually and for two months before calving is dried off—not milked) and pregnant heifers, or young female cows. In the third group are the yearling heifers that are not yet old enough to be bred.

After a paddock has been grazed, we move the cows onto the next paddock. The grazed forage is allowed to rest and regrow as the cows rotate through other paddocks.

We normally begin grazing early, just as the grass begins to grow vigorously around the first week of April–even though the forage is short. These first-grazed paddocks will then be regrown and re-grazed at prime stage when we complete the first rotational cycle. The paddocks around the farm then fall into a rotational sequence of grazing and re-grazing.

In spring and early summer the grass grows fast, so we rotate the cows back to each paddock in about 20 or 30 days. Often, the pasture forage grows faster than the cows can graze it, so we harvest some of the paddocks as hay for winter feed As the summer gets hotter and drier, the forage growth rate slows, so we slow the rotation down to 40 days or more of recovery time before re-grazing a paddock. That allows the forage plants adequate time to recover and regrow from the previous grazing episode.

Without adequate recovery time, the forage plants would become less productive over time, and eventually some plants would die from the stress of overgrazing. Good management promotes greater pasture productivity, higher nutritional value of pasture forages and greater diversity of forages—and ultimately more milk production per acre of land. Good pasture management also improves soil quality and fertility, helps protect water quality, and contributes to wildlife habitat. In short, we are rebuilding our farm's ecological capital, and improving the long-term productivity of the land.

We are also able to continue grazing after the growing season ends in fall. We defer grazing some pasture areas in August and allow the forage there to grow and "stockpile" until the growing season ends in October. We then graze those areas throughout November and even into December, until the grass becomes covered by snow. By starting grazing as early as possible in the spring and stockpiling forage for grazing past the end of the growing season, we are able to lengthen the time the cows can graze each year to about eight months, and we reduce the amount of hay we need to harvest mechanically and store for winter feeding.

The grazing system that we use on our dairy farm mimics the prairie grass/bison ecology that contributed to building the Midwest's deep, fertile prairie soils. However, unlike the roaming bison herds, we manage where our cows graze at all times, which allows us to optimize forage productivity and utilization and to maximize the rebuilding of soil as ecological capital.

Management is important. If paddocks are allowed too much recovery time, the plants will become overly mature and will lose nutritional value. With too little recovery time, some plant species will not recover fully and will die, reducing pasture productivity and diversity. Under good management, plant diversity is maintained or increased and soil fertility is continuously regenerated.

Over time, we have learned many simple management techniques that increase productivity and reduce energy needs. For example, we try to maintain our pasture forage mix at about half grass and half clover. Clover is an important component of the pasture forage mix because clover is a legume that fixes nitrogen from the air into a form the plant can use to make protein. Nitrogen fixation by legumes is done through a symbiotic relationship between legume plants and rhizobia bacteria that inhabit the plants' roots. With adequate clover in our pastures, we do not need to add any nitrogen fertilizer. For that matter, we do not add any fertilizers to our pastures beyond what is recycled back to the soil in the manure—which has been enough to not only replenish but to rebuild the soil fertility in the pastures.

Sometimes paddocks will become sod-bound as grasses begin to predominate and clovers fade from the pasture. To reverse that trend, we allow the cows to graze the sod-bound pastures during rainy times. That will allow the cows' hooves to cut through and open up the sod, allowing more spaces for clover plants to take root and grow. If clover plants begin to dominate over grass in some paddocks, we can allow those paddocks more rest time between grazing episodes, which will help strengthen the grass plants and increase grass presence in the paddock.

The design of a grass-based system is key to making it efficient to manage. We built rock-surfaced lanes throughout the pasture areas to allow the cows to walk from the milking parlor to all paddocks without making mud or causing soil erosion. We use a solar-powered watering system for the cows in the pasture and have water tanks in all paddocks so cows always have access to water.

The solar-powered watering system is set up with an array of solar photovoltaic panels on the edge of a farm pond. The solar panels power a pump in the pond that sends water to a 4,000-gallon tank located on top of the highest hill on the farm. The water then gravity flows from the large tank through an underground pipe system to small drinking tanks on all 60 paddocks on the farm. This system saves us about $150 per month in costs we had previously paid for water from the rural water system. (Farmers in southern Iowa get most of their water from rural water systems because southern Iowa lacks abundant groundwater).

We are working on other renewable energy systems for our farm. In 2010, we installed solar hot water panels on the roof of our dairy processing plant to heat water for our processing plant and milking parlor. We are now making plans to install a wind turbine to produce electricity for the farm.

In 2010, we were also able to purchase an additional 220 acres of land immediately adjacent to our dairy. Now, with about 450 total acres on the farm, we are looking to make our dairy as self-sufficient as possible for our approximately 160 head of dairy animals. We will be able to expand our operation incrementally as our market grows. We do feed some grain, currently about five to six pounds of grain per milking cow during the summer months when the cows are on pasture, and 10 to 12 pounds of grain during the winter when the cows are on stored feed.

It's estimated that 60 million bison once roamed the prairies and plains of North America, compared with about 95 million cattle in the U.S. today.) The bison traveled in large herds and their behavior contributed to the development of fertile soils. Our dairy cattle are not bison and they do not roam our farm in large herds. Far from it. But bison herds roving the prairie landscape are a useful model we can employ to design animal production systems that are resilient, energy-efficient, biologically diverse and ecologically sound.

Acting in Farmscape

Jim O'Loughlin

I was fortunate to appear in *Farmscape* for three performances, in two different locations, and with various casts. The only constants were Mary Swander's enthusiasm and the knowledge that the script in our hands contained the words of fellow Iowans. But that was enough to make for memorable performances.

I played two simultaneous roles in the performances, Mike, a farmer with 1,700 acres, and Nate, a Monsanto employee. Though the blocking posed some challenges (I would whip a seed cap on and off my head to indicate the change), playing both roles made me think about these two characters side by side. Both are middle-aged men who define themselves by their competence and professionalism, but their experiences have taken them in dramatically different directions.

Mike grew up in a farming family and began his career by renting the land he farmed. Slowly, he built up enough capital to buy an acreage and then expand. Nate also grew up on a family farm, but during the farm crisis of the 1980s decided to get an agriculture degree and moved into the hybrid seed industry. Both men had worked hard to get where they were, and both recognized that they had been fortunate to find a path where others had struggled.

During performances, I often found that the stories of Mike and Nate formed a counterbalance to that of Martin. Martin's story was heartbreaking. The family farm he was forced to give up remained such a vital part of his personality that, when interviewed, he was able to recite from heart a list of the farm equipment he'd had to put up for auction. It was a poignant illustration of loss, and there was no necessary reason why things turned out well for Mike and Nate, but not for Martin.

Importantly, *Farmscape* doesn't ask you to choose among these individuals. None is necessarily representative of contemporary rural life. Rather, the play asks audiences to acknowledge the significance of all the experiences documented. In many ways, it is a drama of testimony, where the characters lay bare their lives, their joys, and their sadnesses.

There are echoes of the Great Depression in *Farmscape*. Similar to theatrical innovations of the 1930s like the Federal Theater's *Living Newspaper* project, *Farmscape* brings the news we're not as likely to hear, because it is not about headlines or huge events. Rather, *Farmscape* tells the stories of those who are trying to hang on to a family farm or working in unseen meatpacking jobs or balancing farm work with raising a family. The play details the experiences of farmers whose success has exceeded their expectations, as well as the stories of those who have suffered as centralized agriculture has led to diminished opportunities in rural America.

Much of the play has the feel of stories told around a kitchen table, which in fact is how much of the script's content was obtained. The play drew on language taken from interviews conducted by students working with Mary Swander at Iowa State University. The reality of the dialogue mattered to the performers and to the audience members.

In addition to the camaraderie of the casts, what was most memorable about the performances was the reaction of audiences. Even in a world with a seemingly endless array of entertainment options available on screens, there remains something unique about attending a live, local performance. Seeing people you know on stage enact a play about recognizable people can be a powerful experience.

After performances, many people who attended the play wanted to talk to the actors. They wanted to share their own experiences, in what always felt to me like a continuation of *Farmscape* by other means. One friend of mine who attended a performance grew up on a dairy farm, but he had never spoken in as much detail about that time in his life as he did following the play. Many farmers at one performance wanted to sing the praises of the John Deere 4020, the tractor called "sexy" in one of the biggest laugh lines of the play. I remember a couple coming up afterward and acknowledging the truth of that claim. In fact, the wife said she wondered sometimes whether she or the 4020 was more attractive to her husband.

Other significant parts of this process took time to sink in. It wasn't until the third performance, and my second on stage with Jaime Reyes (who played himself in two performances) before I realized that I had been to his bed and breakfast at a reception for my sister and brother-in-law. It's a beautiful location, as is shown in the slideshow that is part of the play. I had watched those slides and heard Jaime speak of the location many, many times during rehearsals and performances. I felt like an idiot not to realize until the very end of the process that I had my own connection to this place. I'm glad I eventually had the opportunity to share that with Jaime.

The fluid, open manner in which Mary Swander arranged for *Farm-scape* to be performed, allows the text to grow and evolve. I believe it becomes something different each time it is staged. This is appropriate because one of the things this play is about is change, the change that has come to rural America and the ways in which people have benefited from, suffered from, and adapted to that change. In this regard, those of us who acted in the play are like those from whom the play's script was drawn. All of us were trying to make our way under shifting conditions and circumstances.

F is for...

Laura Sweeney

I was different.

One of Ames, Iowa's best-kept secrets is the patio on the back of the West Street Deli. It's a lovely garden, a sanctuary, where I can relax, enjoy the wildflowers, and hear the sounds of nature. When I need a break from my drab, clinical, cubicle world, I go there. Accompanied by an iced tea, egg salad on rye, no spreads—I hang out, read, edit.

One early September afternoon, I was editing the essays for my Choralaires project. I'd spent the summer wrapping up the *Celebrating Extraordinary-Ordinary: Choralaires* documentary, a community arts project about a rural Iowa vocal arts group I'd sung in for about 10 years. Since my graduation had been put on hold I could spearhead this project. The Choralaires wanted the documentary out for their 25th anniversary that year. I'd focused on the video instead of graduation.

I'd intended to graduate in 2006, but a near fatal car accident disrupted my life, and plans for defending my thesis were sidelined as I focused on adjusting to life post-accident. As I worked, I recalled a recent visit to the orthopedic surgeon who had been attending me at McFarland Clinic. Even though my back injury wasn't a surgical issue, and there wasn't much he could do to resolve my pain, I appreciated my doctor's attempt to be reassuring, to show empathy and support. I even appreciated his sense of humor. He understood the dilemma I was in, being advised to limit my stress so I could heal, while at the same time facing the demands to complete graduate school. He joked with me once about the insistent "F word:" when was I going to Finish? But his advice was unnerving. There was no definite prognosis, and I was left to my own devices to learn how to live day-to-day with a chronic injury.

Having exhausted the medical options, I was in search of an alternative route to wellness. I wasn't in the Iowa State University MFA Creative Writing Program, but my professors had consistently encouraged me—I was a good writer who should develop her craft. I was mulling over the possibility of enrolling in Mary Swander's English 557 course: a special topics creative writing class. The Choralaires project had been a good distraction as I was going through physical therapy, and perhaps Swander's class would also be therapeutic.

I had bought and read Swander's book *The Desert Pilgrim* to get a sense of the kind of person she was, and whether she might understand my predicament. I discovered a deeply spiritual writer, someone who had overcome multiple health challenges by writing her way to wellness. If the book was any indication, she was a woman of faith and integrity, committed to her healing journey and to connecting with community to overcome her pain and isolation. As I edited my essays, I realized I could benefit from a strong female protagonist who might inspire me to work through my health issues by developing myself as a writer.

Swander's course wasn't on my Program of Study, though, and wasn't going to get me any closer to graduation. It might even sideline my thesis. Once again, I'd be faced with the "F word" by my committee: when was I going to Finish? Why dedicate myself to another project and community if it jeopardized finishing my degree?

I believe in interdisciplinary work and projects, and I believed that, if it were my calling to participate, in the end the formal academic aspects would work themselves out. I was different. The western model wasn't working for me, and I needed to think holistically, needed to cope with the ambiguity. My healing would take time, patience, and perseverance. And it would take a community.

Later, after I had enrolled in the course, I returned to the West Street Deli patio to edit Daphne's monologue for *Farmscape*. I was inspired by how Daphne's life—like Swander's—interweaves culture, arts, education, and ecology for well-being. And I admired the entrepreneurial spirit that Daphne shares with her husband Jaime. Their colorful estate presented such a contrast to the windowless graduate office I worked in. I'd covered my walls with Guatemalan textiles and inspirational sticky notes—"Two things you got to do is sing in your local community and communicate to the rest of the world." Neither the bright fabrics, nor Pete Seeger's words did much to lift my spirits.

At the time, I didn't admit to myself or to Daphne or Mary Swander that I needed role models of women going against the status quo. My accident and injury had suddenly altered my identity as a woman. In the face of challenges, I needed to be shown the possibility of doing things differently, and still have a fulfilling and healthier life.

Later on, through the *Farmscape* performance, I would come to respect Lonna's voice as well, as she so poignantly expressed how she felt about her organic garden. All things need to grow, need to be nurtured, in their own way. Each thing has its own tastes, whether cool, or hot, and we need to take each living thing on its own terms.

There is no one right way; there is no one right treatment; everything is different.

I was different.

My own health challenges and range of emotions allowed me to appreciate the feistiness of Daphne and Lonna. They persevered in the face of unjust circumstances. Even though I didn't grow up on a farm and didn't have direct experience with the land, I could relate to Kristi's bitterness about her lack of choice, her sense of helplessness and frustration, and the determination to do what she had to do to hold on. I was in that space—struggling to hold on.

I was in a place where I needed to renew myself. *Farmscape* gave me an outlet to play, expand, and experiment with other aspects of my identity. It gave me a route back to the self I felt was getting lost in the academy and medical institutions. I admired these women who wouldn't allow their voices to be taken from them despite their difficult circumstances. I wanted my voice back. I wanted my body back and to re-vision myself. I needed the gumption these women modeled.

I needed the gumption to put my inhibitions aside and pick up a pen and learn to write.

Like these *Farmscape* women, I learned that resilience is doing the best we can to heal from whatever damage has been done, damage that has been out of our control, has not received the recognition it deserves, damage that might have been avoided if not for negligence, that should be remedied. Even though we have this idea that we can always overcome, resilience is about learning to live with what is, on a day-to-day basis. Resilience is not learning how to wave a magic wand so the problem goes away. Resilience starts in the small spaces close to home. Small spaces, like writing.

When we think of writing, what comes to mind is science fiction or fantasy or Hollywood—not writing as a source of healing. But writing can be an important vehicle for reconstructing us, and our communities. When we have the courage to do the hard work of diving into the wreck, exploring the submerged contents, and scouring the treasures that lie beneath, writing can enact a powerful healing that not only helps us reconnect with self, but with others. *Farmscape* serves as an important example of the benefits and impact of writing to wellness.

Four years after our initial *Farmscape* performance, my community work has evolved towards coaching others how to write their way to wellness. I still relax in the West Street Deli patio, accompanied by an iced tea, egg salad on rye, no spreads, and enjoy the wildflowers and the sounds of nature while I search for narrative excerpts that might help another woman find her voice. We need new images, new stories, stories told by women about women, stories told about communities by communities, stories for empowerment. I'm still working on the "F word," though, but this summer it will take on a new meaning: Finished.

Agriculture As the Mother of Arts

Gene Logsdon

A lifetime of farming and writing about farming has convinced me that art is the indispensable nurturer of agriculture, but only recently have I come to see that agriculture is also the mother of art. You can't have one without the other, or when you do, nothing good can come from it in the long run. Farming is a dance, a song, a poem, a painting, a photograph, all celebrating the embrace of humanity with the rest of the natural world.

But much that goes by the name of art is an attempt to deny or ignore nature, and much that goes by the name of farming is a similar denial. The connection between art and nature has been severed. By the same token, agriculture is no longer the culture of the fields but the culture of the factories. The result in either case is the sliding of civilization toward chaos. If nature is essentially chaotic, and it might be for all I know, agriculture is the way we survive. We ferret out and follow the orderly flow of cause and effect that weaves through the unpredictability of the food chain and of the weather. Making order from chaos is the function of art, too. In agriculture, the highest art is building good soil to sustain life.

But there is a critical issue that needs to be addressed in this regard. Artists, breaking away from more traditional "realism" in their work, believe that they can ignore the natural order to expand the possibilities of art. Fine and dandy. We all enjoy playful paintings of purple foxes and blue horses if only because we know foxes aren't purple and horses aren't blue. The pinch comes when society sees so many portrayals of purple foxes that it begins to think maybe purple foxes actually exist. When the misshapen world of Picasso is constantly held up to us as high art, we are tempted to think that such misshapenness is the natural order. Similarly, after one sees enough huge tractors tearing up the landscape, and enough gene manipulation rearranging the codification of life, and enough chemical mist fogging up the food chain, we are tempted to think that John Deere and Monsanto are the natural order.

This collision between the real and the abstract in art, or between nature and modern food production becomes something more than just precious pity-pat among philosophers. It requires definition which in turn requires authority. Whose definitions are correct? When a self-proclaimed artist welds three pieces of junk metal together and calls it art, and the art academy displays it on the lawn in front of its classrooms thereby lending its authority to the artist's claim, does that make it art? If I say that it is still junk, who is correct? Who is the Lord God Almighty who can say when art becomes artifice, when agriculture becomes artificiality?

I wrote *The Mother of All Arts: Agrarianism and the Creative Impulse* in 2007, trying to make these same points about art and agriculture. The reaction in art circles was mostly bristling or belittling. The idea that agriculture might be the seedbed of art as well as corn and soy-beans did not sit well with the Lordgodalmighties of art officialdom. Such a notion threatened the art world's authority and their sense of superiority over those of us whom they have not properly accredited as critics. I was actually told once to stick to growing corn and let the more intelligent ranks in society rule on art.

That's when the debate becomes more than just some bland or earnest discussion over whether abstractionist Pablo Picasso or realist Andrew Wyeth has the better approach. The argument turns to a matter of cultural prejudice. When artists, or more often the merchants of art, pretend that misshapen depictions of reality *are* reality, they join the brotherhood of farming industrialists who pretend that their technology supersedes nature's ways. Society allows such arrogance because it has canonized institutional education and institutional wealth with authority not unlike papal infallibility.

What we are dealing with, as Mary Swander's *Farmscape* suggests in its own subtle and subdued way, is a life and death struggle between the defenseless, mute voice of nature and the formalized, lettered language of institutional intelligence. When I insist that three pieces of junk iron welded together is still junk, I am proving that I do not understand these higher matters of art and technology. Like Swander's farmers I am a mere peasant in the fields of food production, which means that I am a pawn on the artificial field of a chess board to be pushed about at will by the self-proclaimed powers that be.

Farmers of all ranks and ages throughout history have proclaimed the allegiance of art and agriculture in nature. They have never been content just to grow food. They have always wanted to beautify their work even as nature does. For instance, farmers know, as thousands of artists and photographers attest, that corn shocks in stately rows across the fields look beautiful. Farmers understand that shocked corn is still the most environmentally sustainable and economic way to dry corn. The Amish still stubbornly farm this way. In the shock, nature dries the corn, not fossil fuels. The visual beauty of the shocks reflects the deeper beauty of the ecological function. All farmers know artistic expression even though they think they can no longer practice it. How many of them have looked me in the eye in agreement and then said: "I will change the minute everyone else does." Even millionaire grain farmers with thousands of acres of land are pawns and peasants, pushed around by a money interest economy that has severed itself from the true economy of nature.

Beauty may be in the eye of the beholder but how many classic paintings of metal confinement hog buildings have you seen lately compared to stone and wood Pennsylvania Dutch barns? Only the Amish can afford to build the latter anymore. The most telling sign of the decline of good farming today is its diminishing farmstead beauty.

Those of us practicing pasture farming or rotational grazing invariably mow each pasture plot at the end of its grazing rotation. The experts used to tell us such mowing was unnecessary. We ignored them because mowing made the field look pretty. Then we learned that depending on the haphazard grazing of animals to keep the weeds in check doesn't work so well in the long run. Mowing along with grazing is essential to weed control. The pretty way turned out to be the right way. I like to think that is almost always true, peasant that

I am. Beauty is always more than skin deep. Form follows function and good form follows good function. Sustainable farming by its very nature is beautiful and artful. When artists are again drawn to pastoral landscapes for inspiration, then we will know that good farming is on the rise again.

Contributors:

Leigh Adcock is executive director of Women, Food, and Agriculture Network (http://www.wfan.org), a non-profit organization created in 1997 that provides information, networking, and leadership development opportunities for women engaged in sustainable agriculture and food systems development. She grew up on a farm in northwest Iowa, and has worked in television, radio, magazines, newspapers, and public relations. She currently lives on an acreage near Ames, Iowa, with her husband and two teenaged sons.

Jason Arbogast has a BA in English and a BA in Elementary Education. He recently received his MFA in Creative Writing and the Environment from Iowa State University. He was one of the original playwrights of *Farmscape*. He has over ten years of teaching experience, including teaching middle school Language Arts for four years in inner city schools, and composition courses at Iowa State University for two years. Among Arbogast's many publications is his short story collection, *Lost and Found in Kalamazoo*.

Frederick Kirschenmann is Distinguished Fellow at the Leopold Center for Sustainable Agriculture at Iowa State University, and President of the Stone Barns Center for Food and Agriculture in Picontico Hills, New York. He is the author of *Cultivating and Ecological Conscience: Essays From a Farmer Philosopher*.

Mary Klotzbach received her BS at the State University of Iowa with graduate classes in education at SUI and UNI. She taught fifth grade for two years before the birth of her four children in Independence, Iowa, where her husband became a District Court Judge. She has enjoyed being active on the local school board, in the Presbyterian Church, PEO, Ladies Literary, LACES, Democratic party, Homeless Council, and Red Hats.

Anna Lappé is a widely respected author and educator, renowned for her work as a sustainable food advocate. The co-author or author of three books and the contributing author to nine others, Anna's work has been widely translated internationally and featured in *The New York Times, Gourmet, O: The Oprah Magazine*, among many other outlets. Named one of *Time*'s "eco" Who's-Who, Anna is a founding principal of the Small Planet Institute and has for more than a decade been a key force in the growing international movement for sustainability and justice in the food chain.

Gene Logsdon is an American man of letters, cultural and economic critic, and farmer. He is a prolific author of essays, novels, and nonfiction books about agrarian issues, ideals, and techniques. Gene Logsdon farms in Upper Sandusky, Ohio. He has written many books and hundreds of articles for numerous publications including *New Farm, Mother Jones, Orion, Utne Reader,* and *Organic Gardening.*

J. Harley McIlrath was a cast member in the 2009 production of *Farmscape* in Grinnell, Iowa. His fiction has appeared in various publications, including the *North American Review, The Seneca Review*, and *The Wapsipinicon Almanac. Possum Trot,* McIlrath's collection of short fiction set in rural Iowa, is available from the Ice Cube Press.

Jim O'Loughlin is an Associate Professor in the Department of Languages & Literature at the University of Northern Iowa. He teaches classes in American literature, creative writing, and professional writing.

Claudia Prado-Meza is a PhD candidate in Sustainable Agriculture at Iowa State University with a certificate in social justice. She has a bachelor's degree in economics from the Universidad de Colima.

Vicki Simpson is the Development Coordinator at the Hearst Center for the Arts in Cedar Falls, Iowa. She has worked for many years in public relations.

Mary Swander is Poet Laureate of Iowa and Distinguished Professor at Iowa State University, and wrote *Farmscape* with her students in the Creative Writing and Environment program in 2007. Swander, who has written extensively on gardening and agriculture, is the author of 12 books and numerous plays and articles—including *The Girls on the Roof* and *The Desert Pilgrim*.

Laura Sweeney received her master's degree in Interdisciplinary Social Science and Public Administration at Iowa State University. She was one of the original playwrights of *Farmscape*. Laura has over ten years of arts-based research and teaching experience which includes documentary, photo-essay, and ethnographic writing for clients as diverse as the U.S. Department of Energy/Ames Laboratory Environmental Management Program, the ISU Women in Science and Engineering Archives, and The Choralaires, a rural Iowa vocal arts group.

Francis Thicke and his wife Susan own and operate a grass-based organic dairy near Fairfield, Iowa. They process their milk on the farm and market their dairy products through local grocery stores and restaurants. Thicke has a PhD in soil fertility and has served as National Program Leader for Soil Science for the USDA-Extension Service in Washington, DC.

 The Ice Cube Press began publishing in 1993 to focus on how to live with the natural world and to better understand how people can best live together in the communities they share and inhabit. Using the literary arts to explore life and experiences in the heartland of the United States we have been recognized by a number of well-known writers including: Gary Snyder, Wes Jackson, Patricia Hampl, Greg Brown, Jim Harrison, Annie Dillard, Ken Burns, Kathleen Norris, Janisse Ray, Alison Deming, Richard Rhodes, Michael Pollan, and Barry Lopez. We've published a number of well-known authors including: Mary Swander, Jim Heynen, Mary Pipher, Bill Holm, Connie Mutel, John T. Price, Carol Bly, Marvin Bell, Debra Marquart, Ted Kooser, Stephanie Mills, Bill McKibben, Gene Logsdon, Anna Lappé, Frederick Kirschenmann, and Paul Gruchow. We have won several publishing awards over the last nineteen years, had reviews in regional and national publications and participated at regional educational events. Check out our books at our web site, join our facebook group, visit booksellers, museum shops, or any place you can find good books and discover why we continue striving to "hear the other side."

Ice Cube Press, LLC (eſt. 1993)
205 N. Front Street
North Liberty, Iowa 52317-9302
ſteve@icecubepress.com
www.icecubepress.com
@icecubepress on twitter

Making good earth, sowing good life
all together with
Fenna Marie & Laura Lee